Memoirs of the American Mathematical Society
Number 331

Ron C. Blei

Fractional dimensions and bounded fractional forms

Published by the
AMERICAN MATHEMATICAL SOCIETY
Providence, Rhode Island, USA

September 1985 · Volume 57 · Number 331 (third of 6 numbers)

MEMOIRS of the American Mathematical Society

SUBMISSION. This journal is designed particularly for long research papers (and groups of cognate papers) in pure and applied mathematics. The papers, in general, are longer than those in the TRANSACTIONS of the American Mathematical Society, with which it shares an editorial committee. Mathematical papers intended for publication in the Memoirs should be addressed to one of the editors:

Ordinary differential equations, partial differential equations and applied mathematics to JOEL A. SMOLLER, Department of Mathematics, University of Michigan, Ann Arbor, MI 48109

Complex and harmonic analysis to LINDA PREISS ROTHSCHILD, Department of Mathematics, University of California at San Diego, La Jolla, CA 92093

Abstract analysis to WILLIAM B. JOHNSON, Department of Mathematics, Texas A&M University, College Station, TX 77843-3368

Classical analysis to PETER W. JONES, Department of Mathematics, Yale University, New Haven, CT 06520

Algebra, algebraic geometry and number theory to LANCE W. SMALL, Department of Mathematics, University of California at San Diego, La Jolla, CA 92093

Logic, set theory and general topology to KENNETH KUNEN, Department of Mathematics, University of Wisconsin, Madison, WI 53706

Topology to WALTER D. NEUMANN, Mathematical Sciences Research Institute, 2223 Fulton St., Berkeley, CA 94720

Global analysis and differential geometry to TILLA KLOTZ MILNOR, Department of Mathematics, Hill Center, Rutgers University, New Brunswick, NJ 08903

Probability and statistics to DONALD L. BURKHOLDER, Department of Mathematics, University of Illinois, Urbana, IL 61801

Combinatorics and number theory to RONALD GRAHAM, Mathematical Sciences Research Center, AT&T Bell Laboratories, 600 Mountain Avenue, Murray Hill, NJ 07974

All other communications to the editors should be addressed to the Managing Editor, R. O. WELLS, JR., Department of Mathematics, Rice University, Houston, TX 77251

PREPARATION OF COPY. Memoirs are printed by photo-offset from camera-ready copy prepared by the authors. Prospective authors are encouraged to request a booklet giving detailed instructions regarding reproduction copy. Write to Editorial Office, American Mathematical Society, Box 6248, Providence, RI 02940. For general instructions, see last page of Memoir.

SUBSCRIPTION INFORMATION. The 1985 subscription begins with Number 314 and consists of six mailings, each containing one or more numbers. Subscription prices for 1985 are $188 list, $150 institutional member. A late charge of 10% of the subscription price will be imposed on orders received from nonmembers after January 1 of the subscription year. Subscribers outside the United States and India must pay a postage surcharge of $10; subscribers in India must pay a postage surcharge of $15. Each number may be ordered separately; *please specify number* when ordering an individual number. For prices and titles of recently released numbers, see the New Publications sections of the NOTICES of the American Mathematical Society.

BACK NUMBER INFORMATION. For back issues see the AMS Catalogue of Publications.

Subscriptions and orders for publications of the American Mathematical Society should be addressed to American Mathematical Society, Box 1571, Annex Station, Providence, RI 02901-1571. *All orders must be accompanied by payment.* Other correspondence should be addressed to Box 6248, Providence, RI 02940.

MEMOIRS of the American Mathematical Society (ISSN 0065-9266) is published bimonthly (each volume consisting usually of more than one number) by the American Mathematical Society at 201 Charles Street, Providence, Rhode Island 02904. Second Class postage paid at Providence, Rhode Island 02940. Postmaster: Send address changes to Memoirs of the American Mathematical Society, American Mathematical Society, Box 6248, Providence, RI 02940.

The paper used in this journal is acid-free and falls within the guidelines established to ensure permanence and durability.

TABLE OF CONTENTS

iii

ABSTRACT

Under definitions of combinatorial dimension, fractional Cartesian products and bounded fractional forms on $C_o(\mathbb{N})$, the main result of the first part of the paper is a precise relationship between ℓ^p-norms of restrictions of bounded fractional forms and combinatorial dimensions of subsets of $\mathbb{N}^J$. This extends classical inequalities due to Littlewood and Johnson & Woodward regarding bounded multi-linear forms on $C_o(\mathbb{N})$.

In a framework of multi-linear measure theory, under definitions of fractional Fréchet pseudomeasures (measure theoretic analogues of bounded fractional forms) and dimension of sets in a Borel measurable setting, the main results of the second part of the paper are

(i)   a 'fractional-linear' Riesz Representation Theorem, extending
       (the 'bilinear') work of Fréchet and Morse & Transue,

and

(ii)   relationships between variations of Fréchet pseudomeasures
       and dimensions of subsets of $[0,1]^J$.

1980 Mathematics Subject Classification. Primary 26D15, 46G99; Secondary 05A99, 28A35.

Key words and phrases. Combinatorial dimension, fractional Cartesian products, bounded fractional forms, Fréchet pseudomeasures.

Library of Congress Cataloging-in-Publication Data

Blei, R. C. (Ron C.)
    Fractional dimensions and bounded fractional forms.

    (Memoirs of the American Mathematical Society, ISSN 0065-9266;
no. 331)
        Bibliography: p.
        1. Functions of real variables. 2. Inequalities (Mathematics)
3. Measure theory. 4. Combinatorial analysis. I. Title. II. Series.
QA3.A57   no. 331   [QA331.5]   510s   [515.8]   85-13512
ISBN 0-8218-2332-9

0. INTRODUCTION

So called fractional Cartesian products were designed in [3] in order
to fill analytic and combinatorial gaps between ordinary Cartesian products of spectral sets. Leading to a subsequent paper [4], the constructions of [3] suggested the measurement of a 'continuous' parameter
attached to a subset of a Cartesian product, designated as *combinatorial
dimension*. Although cast in a particular setting of harmonic analysis on
discrete groups, both papers [3] and [4] displayed, in effect, a precise
interaction between analytic properties of bounded multilinear forms on c_o
and the combinatorial dimension of subsets of $\mathbb{N}^J$. Following standard
terminology, we say that an array of scalars (a J-tensor)

$$a = (a_{i_1 \ldots i_J})_{i_1, \ldots, i_J \in \mathbb{N}}$$

is a bounded J-linear form on c_o if

(0.1) $\displaystyle \left| \sum_{i_1, \ldots, i_J = 1}^{N} a_{i_1 \ldots i_J} s_1(i_1) \cdots s_J(i_J) \right| \le \eta \sup_i |s_1(i)| \cdots \sup_i |s_J(i)|$

for all sequences of scalars $(s_1(i))_{i \in \mathbb{N}}, \ldots, (s_J(i))_{i \in \mathbb{N}}$ and all $N \ge 1$;
the smallest $\eta \ge 0$ for which (0.1) holds is the norm of the bounded
J-linear form, denoted by $\|a\|$. Given $F \subset \mathbb{N}^J$, we consider the
ℓ^p-norms of restrictions to F of J-tensors, denoted by

$$\|a|_F\|_p = \left(\sum_{(i_1, \ldots, i_J) \in F} |a_{i_1 \ldots i_J}|^p \right)^{1/p}.$$

Under a definition of the combinatorial dimension of $F \subset \mathbb{N}^J$, denoted by
dimF, one of the main results in [4] was

Theorem 0.1 (Th. 5.2[4])

Let J be a positive integer, and let F be an arbitrary subset of
$\mathbb{N}^J$. Then,

Received by the editors November 29, 1984.
Research partially supported by NSF grant #MCS 8301659.

1

(0.2) $\inf \{p: \sup_{\substack{a \text{ bdd. J-linear} \\ \text{form, } \|a\|=1}} \|a|_F\|_p < \infty \} = 2/(1 + 1/\dim F)$

(the parameter defined by the left hand side of (0.2) is designated as σ_F).

The case $F = \mathbb{N}^J$ in formula (0.2) $(\dim \mathbb{N}^J = J)$ is classical ([12],
[8]). The "fractional Cartesian products," denoted here by $\mathbb{N}^{J/K}$, were
shown in [3] to satisfy

(0.3) $\sigma_{\mathbb{N}^{J/K}} = 2/(1 + K/J)$, $J \geq K > 0$ arbitrary integers,

thus filling the gap between $\mathbb{N}^J$ and $\mathbb{N}^{J+1}$. The subsequent Theorem 0.1
above was a general statement which in particular, after the observation

$$\dim \mathbb{N}^{J/K} = J/K ,$$

implied (and explained) the equality in (0.3).

In the present work, we do the following:

(i) The notion of fractional Cartesian products, which in [3] led to
examples of sets with prescribed fractional dimensions, here gives rise to
a framework for a general statement extending Theorem 0.1: We consider
here the notion of bounded *fractional* forms on c_o whose ℓ^p-norms we
link, extending (0.2), to a measurement of the combinatorial dimension of
subsets in $\mathbb{N}^J$. These matters are worked out in section 3.

(ii) The preceding papers [3] and [4], and sections 1-3 of this
paper were cast in a 'discrete' setting of a J-fold Cartesian product of a
set devoid of any 'internal' structure. In sections 4 and 5, guided by
Fréchet's point of view in [6] (briefly discussed in the next section), we
recast and extend the 'discrete' notions of sections 1, 2, and 3 in a
general framework of measurable product spaces.

We proceed now to a more detailed account of what is done here. In
section 1, as well as formalizing necessary preliminaries, we describe
some of the classical notions that led to the present paper. The historical line is essentially this: Fréchet's 1915 work in the 'continuous'
setting $[0,1] \times [0,1]$, characterizing the bounded bilinear functionals on
$C([0,1])$ ([6]), motivated Littlewood's 1930 work in the 'discrete' setting

$\mathbb{N} \times \mathbb{N}$, establishing two-dimensional inequalities regarding bounded bi-linear forms on $c_0(\mathbb{N})$ ([12]); much later (1974), following Littlewood's 'two-dimensional' ideas, the multilinear version of these inequalities was worked out in [8]. In this section, we reprove the J-linear Littlewood inequalities by induction on J, starting with the trivial case J = 1 (Theorem 1.2). The proof given here yields an improved growth of constants which exhibits a surprising dependence on the dimension J (Remark 1.5). In section 1, 'dimension' is still a positive integer.

In section 2, we formalize and explain the concepts of 'combinatorial dimension' and 'fractional Cartesian products.' X will denote here an infinite set without structure. The combinatorial dimension of $F \subset X^J$, described briefly, is a measurement of interdependencies between the J canonical projections *restricted to* F into the respective 'coordinate axes (Definition 2.1). Along the same line, the idea of fractional Cartesian products is based essentially on the observation that points in X^J, generically determined by the J 'independent' projections from X^J onto X, can be described also in terms of 'interdependent' projections from X^J onto prescribed 'coordinate hyperplanes' (Definition 2.3). Measurements of combinatorial dimensions of the fractional Cartesian products defined in this section (Theorem 2.5 and Corollary 2.6) are closely related to 'measure theoretic' isoperimetric inequalities (Theorem 2.8).

In section 3, we extend Theorem 0.1 in the framework of fractional Cartesian products. Under the notion of bounded *fractional* linear forms on c_0, defined by (3.5), we establish

Theorem 0.2 (Theorem 3.1)

Let $J \geq K > 0$ be arbitrary integers, and let Λ be an arbitrary infinite subset of $\mathbb{N}^J$. Then,

$$\inf \left\{ p : \sup_{\substack{a \text{ bdd. } J/K\text{-linear} \\ \text{form, } \|a\|=1}} \| a |_\Lambda \|_p < \infty \right\} = \max \{1, 2/(1 + K/\dim\Lambda)\}.$$

Theorem 0.1 (Th. 5.2 [4]) is the instance K = 1 in Theorem 0.2 above. The methods of proof in section 3 are essentially combinatorial; Khintchin's inequality and the fundamental Kahane-Salem-Zygmund estimates are the only analytic tools required here. Indeed, while the arguments in [4], establishing Theorem 0.1, were based on Pisier's results in [17] regarding multipliers from L^2 into the 'exponential-square' class, the methods of the present paper depend only on L^1-L^2 inequalities.[*]

In section 4 we define and study the measure-theoretic analogues of bounded multilinear forms on c_o. For the purpose of the brief account here, let X be a measurable space and let X^J denote its usual measurable J-fold product. Fix integers $J \geq K > 0$ and let μ be a scalar valued function defined on the measurable rectangles in X^J. We say that μ is an $F_{J/K}$-pseudomeasure if, fixing *any* J-K coordinates, we have that μ, defined on the remaining K coordinates, is extendible to a complex measure on X^K (Definitions 4.1 and 4.6). The space of $F_{J/K}$-pseudomeasures is denoted by $F_{J/K}$ and appropriately normed by the $F_{J/K}$-variation ((4.3), Theorem 4.3, Theorem 4.8). Integration with respect to $F_{J/K}$-pseudomeasures is obtained by induction following the case $J = 1$, which is usual Lebesgue integration with respect to complex measures (Lemma 4.9, Proposition 4.10, (4.32), Proposition 4.11). When X is a locally compact Hausdorff space, after a definition of a J/K-fold projective tensor product of $C_o(X)$, denoted by $V_{J/K}$ ((4.40), (4.44)), we deduce a 'J/K-linear Riesz Representation Theorem'

Theorem 0.3 (Theorems 4.12 and 4.14)

$$V^*_{J/K} = F_{J/K} .$$

The instance $J = 2$, $K = 1$, and $X = [0,1]$ in the theorem above is Fréchet's characterization of bounded bilinear functionals on C([0,1]) in [6].

In section 5, extending the ideas of section 2 to a Borel measurable framework, we consider the combinatorial dimension of $F \subset [0,1]^J$ relative

---

[*]Some of the ideas underlying Pisier's work [17] can be traced back to Kolmogorov's and Tihomirov's paper on entropy [11]; this suggests a more direct connection between [11] and the present work.

to the Borel field in $[0,1]$, denoted by DimF (Definition 5.1). Given an $F_{J/K}$-pseudomeasure μ on $[0,1]^J$ and a subset F of $[0,1]^J$, we consider the pth-variation of μ over F (defined in (5.9)) and the subsequent 'Hausdorff dimension' of F relative to μ, denoted by H-$\text{Dim}_\mu F$ (defined in (5.11)). The main result is

Theorem 0.4 (Theorems 5.6 and 5.7)

Let $J \geq K > 0$ be arbitrary integers, and let F be an arbitrary subset of $[0,1]^J$.

(a) For all $\mu \varepsilon F_{J/K}$

$$H\text{-Dim}_\mu F \leq \max \{1, \ 2/(1 + K/\text{DimF})\} \ .$$

(b) If $\text{DimF} = \text{dimF} \geq 1$ then there are $\mu \varepsilon F_{J/K}$ for which

$$H\text{-Dim}_\mu F = \max \{1, \ 2/(1 + K/\text{DimF})\} \ .$$

The "fractionally-linear" issues treated here require, by their very nature, "fractionally-linear" notation; the reader is asked to be patient and forgiving. The notation of this paper is recalled and developed as progress is made. A common recurrence here is the following: Let $f_1, \ldots, f_J$ be scalar valued functions on $X_1, \ldots, X_J$, respectively. $f_1 \otimes \cdots \otimes f_J$ denotes the function on $X_1 \times \cdots \times X_J$ defined by

$$f_1 \otimes \cdots \otimes f_J(x_1, \ldots, x_J) = f_1(x_1) \cdots f_J(x_J), \quad x_1 \varepsilon X_1, \ldots, x_J \varepsilon X_J \ .$$

To avoid splitting arguments into real and imaginary parts, all the work in this paper is done over the field of real numbers; modulo numerical constants, all that is done holds as well with complex scalars.

The paper was written during my visit to the University of British Columbia in the academic year 1983-84 which was both enjoyable and stimulating. I want to thank particularly John Fournier and Ed Granirer whose N.S.E.R.C. grants made my visit at U.B.C. possible.

1. FRÉCHET'S BOUNDED BILINEAR FUNCTIONALS AND LITTLEWOOD'S BOUNDED BILINEAR FORMS

In a 1915 paper "Sur les fonctionnelles bilinéaires [6]," M. Fréchet extended naturally the Riesz Representation Theorem [18] whose statement is essentially the following:

ν is a bounded *linear* functional on $C([0,1])$, the space of continuous functions on $[0,1]$, if and only if there is a function of bounded total variation $\phi = \phi_\nu$ so that

$$\nu(f) = \int_{[0,1]} f(x)\,d\phi(x) \qquad \text{(Riemann-Stieltjes integral)}$$

for all $f \in C([0,1])$.

The corresponding 'two-dimensional' problem of characterizing bounded *bilinear* functionals on $C([0,1])$ was solved in [6] via an extension of the usual notion of bounded variation: Let ϕ be a function of two variables on $[0,1] \times [0,1] = [0,1]^2$. Given two partitions of $[0,1]$

$$\pi = \{0 = x_0 \le x_1 \cdots \le x_{K-1} \le x_K = 1\}$$

and

$$\tau = \{0 = y_0 \le y_1 \cdots \le y_{N-1} \le y_N = 1\},$$

write

$$\Delta_{kn}^{\pi\tau}\phi = \phi(x_{k+1}, y_{n+1}) - \phi(x_k, y_{n+1}) + \phi(x_k, y_n) - \phi(x_{k+1}, y_n),$$

and define the Fréchet variation of ϕ

$$(1.1) \qquad \Delta_F\phi = \sup_{\pi,\tau} \sup_{\substack{\varepsilon_k = \pm 1 \\ \delta_n = \pm 1}} \left| \sum_{k,n} (\Delta_{kn}^{\pi\tau}\phi)\varepsilon_k\delta_n \right|.$$

To place $\Delta_F \phi$ in its proper context, note that the usual total variation of ϕ (in the sense of Vitali) is given by

$$\Delta_V \phi = \sup_{\pi,\tau} \sum_{k,n} |\Delta_{kn}^{\pi\tau} \phi|$$

$$= \sup_{\pi,\tau} \sup_{\varepsilon_{kn} = \pm 1} |\sum_{k,n} (\Delta_{kn}^{\pi\tau} \phi) \varepsilon_{kn}| .$$

Having thus extended the notion of variation over $[0,1]^2$, Fréchet proved:

ν is a bounded *bilinear* functional on $C([0,1])$ if and only if there is a function $\phi = \phi_\nu$ on $[0,1]^2$ with $\Delta_F \phi < \infty$ so that

$$(1.2) \qquad\qquad \nu(f,g) = \int_{[0,1]^2} f(x) g(y) d\phi(x,y)$$

for all $f, g \in C([0,1])$, where the 'integral' above is an appropriately defined iterated Riemann-Stieltjes integral (see pp. 225-227 in [6]).

At this juncture observe that every bounded linear functional on $C([0,1]^2)$, by its action on $f \otimes g$, $f, g \in C([0,1])$, gives rise to a bounded bilinear functional on $C([0,1])$; indeed, every function ϕ on $[0,1]^2$ with $\Delta_V \phi < \infty$ a fortiori satisfies $\Delta_F \phi < \infty$. The converse, however, is false:[1] At the very outset of his 1930 paper "On bounded bilinear forms in an infinite number of variables [12]," Littlewood noted that the problem of producing an example of a function of two variables with finite Fréchet variation but infinite total variation is equivalent to the problem of producing an array of scalars $(a_{mn})_{m,n \in \mathbb{N}}$ satisfying

$$(1.3) \qquad\qquad |\sum_{m,n} a_{mn} s_m t_n| \leq 1 \quad \text{for all sequences}$$

$$(s_m)_{m \in \mathbb{N}} \quad (t_n)_{n \in \mathbb{N}} \quad \text{in the unit ball of } c_o ,$$

and

$$\sum_{m,n} |a_{mn}| = \infty .$$

---

[1] Whether the converse holds is a natural question which Fréchet did not consider in his paper [6].

(Analogous to Fréchet's bounded bilinear functionals on $C([0,1])$, matrices satisfying (1.3) are precisely the bounded bilinear functionals on $c_0(\mathbb{N}) = c_0$ and are called bounded bilinear forms.) Having rephrased the problem, Littlewood quickly constructed an example based on Hilbert's inequality (p. 164 of [12]), and proceeded to consider a resulting question: Every bounded bilinear form is square summable but not necessarily absolutely summable; is there $1 < p < 2$ so that every array $(a_{mn})_{m,n\varepsilon\mathbb{N}}$ satisfying (1.3) necessarily satisfies

$$\sum_{m,n} |a_{mn}|^p < \infty \ ?$$

The answer given in [12] was this:

All bounded bilinear forms $(a_{mn})_{m,n\varepsilon\mathbb{N}}$ satisfy

(1.4)
$$\sum_m (\sum_n |a_{mn}|^2)^{1/2}, \quad \sum_n (\sum_m |a_{mn}|^2)^{1/2} < \infty$$

which implies

(1.5)
$$\sum_{m,n} |a_{mn}|^{4/3} < \infty .$$

Moreover, the exponent $4/3$ in (1.5) is best possible in the sense that

(1.5)[#] there are bounded bilinear forms $(a_{mn})_{m,n}$ for which

$$\sum_{m,n} |a_{mn}|^p = \infty \quad \text{for every} \quad p < 4/3 .$$

Littlewood's inequalities, stated above in a 'two dimensional' setting, have natural analogues within a multi-dimensional framework which we now describe. Let

$$R = \{r_n\}_{n\varepsilon\mathbb{N}}$$

be the usual Rademacher system realized as a sequence of functions on $[0,1]$:

$$r_n(t) = 1 - 2\varepsilon_n , \quad n\varepsilon\mathbb{N} ,$$

where $t = \sum_{n=1}^{\infty} \varepsilon_n/2^n$ is the binary expansion of $t\varepsilon[0,1]$. Viewed on the

probability space ([0,1], Lebesgue measure), R is a system of statisti-
cally independent symmetric random variables taking values +1 and -1
with probability 1/2. Two basic properties of the Rademacher functions
are these:

$$(1.6) \qquad \sup_{t \varepsilon [0,1]} |\sum_n a_n r_n(t)| = \sum_n |a_n| \ ;$$

$$(1.7) \qquad \kappa \int_{[0,1]} |\sum_n a_n r_n(t)| dt \geq \left(\int_{[0,1]} |\sum_n a_n r_n(t)|^2 dt \right)^{1/2}$$

$$= \left(\sum_n |a_n|^2 \right)^{1/2} ,$$

and

$$(1.7)_p \qquad \kappa_p \left(\int_{[0,1]} |\sum_n a_n r_n(t)|^p \right)^{1/p} \geq \left(\sum_n |a_n|^2 \right)^{1/2} , \quad 1 < p < 2 ,$$

where $(a_n)_{n \varepsilon \mathbb{N}}$ is an arbitrary sequence of scalars, and κ, κ_p denote
the 'best' constants in the respective inequalities above. (1.6) is ob-
tained simply by taking $t \varepsilon [0,1]$ so that $r_n(t) = |a_n|/a_n$, $a_n \neq 0$. The
inequality (1.7) (or $(1.7)_p$, which is equivalent to (1.7) modulo 'best'
constants) is the classical Khintchin inequality, a fundamental probabilistic fact obtained and published in various contexts independently by
several mathematicians during the 1920's and 1930's ([10] is an early
reference). Indeed, in his paper on bounded bilinear forms [12],
Littlewood established (1.7) (with $\kappa \leq \sqrt{3}$) in order to deduce inequality
(1.4). The best constant $\kappa = \sqrt{2}$ was computed in [21]; another proof of
$\kappa = \sqrt{2}$ as well as a complete determination of κ_p, $1 < p < 2$,
appeared in [7].

Moving to higher dimensions, let J be a positive integer and con-
sider the J-fold Cartesian product of $R = \{r_n\}_{n \varepsilon \mathbb{N}}$ canonically realized
as a system of functions on $[0,1]^J$:

$$(1.8) \qquad R^J = \{r_{n_1} \otimes \cdots \otimes r_{n_J} : (n_1, \ldots, n_J) \varepsilon \mathbb{N}^J \},$$

where $r_{n_1} \otimes \cdots \otimes r_{n_J}$, $(n_1, \ldots, n_J) \varepsilon \mathbb{N}^J$, is the function defined by

$$r_{n_1} \otimes \cdots \otimes r_{n_J}(t_1, \ldots, t_J) = r_{n_1}(t_1) \cdots r_{n_J}(t_J), \quad (t_1, \ldots, t_J) \varepsilon [0,1] .$$

We shall deal here with scalar valued functions, a priori assumed integrable with respect to the Lebesgue measure on $[0,1]^J$, in a prescribed norm closure of the linear span of R^J. Such functions, which we call J-linear forms, are uniquely determined by scalar valued J-tensors

$a = (a_{n_1\cdots n_J})_{n_1,\ldots,n_J \epsilon \mathbb{N}}$ defined by

$$a_{n_1\cdots n_J} = \hat{f}(n_1,\ldots,n_J)$$

$$= \int_{[0,1]^J} f(t_1,\ldots,t_J) r_{n_1}(t_1)\cdots r_{n_J}(t_J) dt_1\cdots dt_J ;$$

a J-linear form f is then written uniquely as

$$(1.9) \qquad\qquad f \sim \sum_{n_1,\ldots,n_J} \hat{f}(n_1,\ldots,n_J) r_{n_1}\otimes\cdots\otimes r_{n_J} .$$

The closure in the supremum norm of the linear span of R^J will be denoted by $C_{R^J}([0,1]^J) = C_{R^J}$; elements of C_{R^J} will be called bounded J-linear forms. Observe that the bounded J-linear forms are precisely the bounded J-linear functionals on c_o whose continuous J-linear action is given by

$$\sum_{n_1,\ldots,n_J} \hat{f}(n_1,\ldots,n_J) s_1(n_1)\cdots s_J(n_J), \qquad f \epsilon C_{R^J} , \quad s_1,\ldots,s_J \epsilon c_o ,$$

$$\left| \sum_{n_1,\ldots,n_J} \hat{f}(n_1,\ldots,n_J) s_1(n_1)\cdots s_J(n_J) \right| \leq \|f\|_\infty \|s_1\|_\infty \cdots \|s_J\|_\infty .$$

In this context, (1.6), (1.5) and (1.5)[#] are statements about bounded 1-linear and 2-linear forms respectively.

Proceeding to treat the general multilinear inequalities, we start with a statement of the J-linear Khintchin inequality, easily obtained by induction on J (for example, see Appendix D in [20]):

$$(1.10) \qquad \kappa^J \int_{[0,1]^J} \left| \sum_{n_1,\ldots,n_J} a_{n_1\cdots n_J} r_{n_1}(t_1)\cdots r_{n_J}(t_J) \right| dt_1\cdots dt_J \geq$$

$$\geq \left(\sum_{n_1,\ldots,n_J} |a_{n_1\cdots n_J}|^2 \right)^{1/2}$$

for every J-tensor $(a_{n_1\cdots n_J})_{n_1,\ldots,n_J \epsilon \mathbb{N}}$. Combining (1.6) and (1.10),

we establish below the multilinear version of Littlewood's bilinear inequality (1.4), whose proof is essentially an adaptation of Littlewood's
'bilinear' argument in [12].

Lemma 1.1

For all integers $J \geq 2$ and all $f \in C_{R^J}$,

$$(1.11) \qquad \kappa^{J-1} \|f\|_\infty \geq \sum_{n_J} \left(\sum_{n_1,\ldots,n_{J-1}} |\hat{f}(n_1,\ldots,n_J)|^2 \right)^{1/2} .$$

Proof

$$\|f\|_\infty = \sup_{t_1,\ldots,t_J} \left| \sum_{n_J} \left(\sum_{n_1,\ldots,n_{J-1}} \hat{f}(n_1,\ldots,n_J) r_{n_1}(t_1) \cdots r_{n_{J-1}}(t_{J-1}) \right) r_{n_J}(t_J) \right|$$

$$= \sup_{t_1,\ldots,t_{J-1}} \sum_{n_J} \left| \sum_{n_1,\ldots,n_{J-1}} \hat{f}(n_1,\ldots,n_J) r_{n_1}(t_1) \cdots r_{n_{J-1}}(t_{J-1}) \right|$$

$$\hspace{9cm}\text{(by 1.6)}$$

$$\geq \sum_{n_J} \int_{[0,1]^{J-1}} \left| \sum_{n_1,\ldots,n_{J-1}} \hat{f}(n_1,\ldots,n_J) r_{n_1}(t_1) \cdots r_{n_{J-1}}(t_{J-1}) \right| dt_1 \cdots dt_{J-1}$$

$$\geq \kappa^{-J+1} \sum_{n_J} \left(\sum_{n_1,\ldots,n_{J-1}} |\hat{f}(n_1,\ldots,n_J)|^2 \right)^{1/2} \qquad \text{(by (1.10)) .} \qquad \boxed{x}$$

Theorem 1.2

For all integers $J \geq 1$ and $f \in C_{R^J}$

$$(1.12) \qquad \lambda_J \|f\|_\infty \geq \|\hat{f}\|_{2J/(J+1)} ,$$

where $(\lambda_J)_{J \geq 1}$ is a sequence of positive constants satisfying inductively

$$\lambda_1 = 1 \quad \text{and} \quad \lambda_J \leq (\kappa \cdot \kappa_{2-2/J} \cdot \lambda_{J-1})^{(J-1)/J}, \quad J > 1 .$$

(Notation: $\|\hat{f}\|_q = \left(\sum_{n_1,\ldots,n_J} |\hat{f}(n_1,\ldots,n_J)|^q \right)^{1/q}$; κ and $\kappa_{2-2/J}$ are
the Khintchin constants given in (1.7) and (1.7)$_p$).

Proof (by induction on J)

The case $J = 1$ is (1.6). Let $J > 1$ and assume that (1.12) holds in the case $J - 1$. Given an arbitrary $f \in C_{R^J}$ and applying the induction hypothesis, we obtain

$$\| f \|_\infty =$$

$$= \sup_{t_1,\ldots,t_J \in [0,1]} | \sum_{n_1,\ldots,n_{J-1}} (\sum_{n_J} \hat{f}(n_1,\ldots,n_J) r_{n_J}(t_J)) r_{n_1}(t_1) \cdots r_{n_{J-1}}(t_{J-1}) |$$

$$\geq (\lambda_{J-1})^{-1} (\sup_t \sum_{n_1,\ldots,n_{J-1}} | \sum_{n_J} \hat{f}(n_1,\ldots,n_J) r_{n_J}(t) |^{2(J-1)/J})^{J/2(J-1)} .$$

Therefore,

$$(\lambda_{J-1} \| f \|_\infty)^{2(J-1)/J} \geq \sum_{n_1,\ldots,n_{J-1}} \int_{[0,1]} | \sum_{n_J} \hat{f}(n_1,\ldots,n_J) r_{n_J}(t) |^{2(J-1)/J} dt .$$

An application of Khintchin's inequality $(1.7)_p$ with $p = 2(J-1)/J$ to the inequality above yields

$$(1.13) \qquad (\kappa_p \lambda_{J-1} \| f \|_\infty)^{2(J-1)/J} \geq \sum_{n_1,\ldots,n_{J-1}} (\sum_{n_J} | \hat{f}(n_1,\ldots,n_J) |^2)^{(J-1)/J} .$$

Write

$$(1.14) \qquad \sum_{n_1,\ldots,n_J} | \hat{f}(n_1,\ldots,n_J) |^{2J/(J+1)} =$$

$$= \sum_{n_1,\ldots,n_J} | \hat{f}(n_1,\ldots,n_J) |^{2(J-1)/(J+1)} | \hat{f}(n_1,\ldots,n_J) |^{2/(J+1)} .$$

At the right hand side of (1.14) apply Hölder's inequality with exponents $(J+1)/(J-1)$ and $(J+1)/2$ to $\sum_{n_J}$, and thus obtain

$$(1.15) \qquad \| \hat{f} \|_{2J/(J+1)}^{2J/(J+1)} \leq \sum_{n_1,\ldots,n_{J-1}} (\sum_{n_J} | \hat{f} |^2)^{(J-1)/(J+1)} (\sum_{n_J} | \hat{f} |)^{2/(J+1)} .$$

Apply Hölder's inequality at the right hand side of (1.15) with exponents $(J+1)/J$ and $J+1$ to $\sum_{n_1,\ldots,n_{J-1}}$, and deduce

(1.16) $\|\hat{f}\|_{2J/(J+1)}^{2J/(J+1)} \leq$

$$\leq \Big(\sum_{n_1,\ldots,n_{J-1}} \big(\sum_{n_J} |\hat{f}|^2 \big)^{(J-1)/J} \Big)^{J/(J+1)} \Big(\sum_{n_1,\ldots,n_{J-1}} \big(\sum_{n_J} |\hat{f}| \big)^2 \Big)^{1/(J+1)} \ .$$

Observe now that Lemma 1.1 implies (via Minkowski's inequality)

(1.17) $\big(\kappa^{J-1} \|f\|_\infty \big)^{2/(J+1)} \geq \Big(\sum_{n_1,\ldots,n_{J-1}} \big(\sum_{n_J} |\hat{f}| \big)^2 \Big)^{1/(J+1)} \ .$

Finally, combining (1.13), (1.16) and (1.17), we deduce

$$\|\hat{f}\|_{2J/(J+1)} \leq \big(\kappa \cdot \kappa_{2-2/J} \cdot \lambda_{J-1} \big)^{(J-1)/J} \|f\|_\infty \ .$$

$$\boxed{\text{x}}$$

To establish that the exponents $2J/(J+1)$, $J = 2,\ldots,$ are best possible in Theorem 1.2 we rely on the Kahane-Salem-Zygmund probabilistic estimates, an instance of which is formalized below for later use in the paper. The proof is omitted (see pp. 54-57 in [9]).

Lemma 1.3 (e.g., Theorem 4 on p. 57 of [9])

Let $N > 1$ be an arbitrary integer, and $A_1,\ldots,A_J \subset \mathbb{N}$ be arbitrary finite subsets each of whose cardinality is N. Given any J-tensor supported on $A_1 \times \cdots \times A_J$

$$a = (a_n)_{n \in A_1 \times \cdots \times A_J}$$

there exists a choice of signs $\pm$ so that

(1.18) $\big\| \sum_{\substack{n \in A_1 \times \cdots \times A_J \\ n=(n_1,\ldots,n_J)}} \pm a_n r_{n_1} \otimes \cdots \otimes r_{n_J} \big\|_\infty \leq c \big(N \sum_{n \in A_1 \times \cdots \times A_J} |a_n|^2 \big)^{1/2} \ ,$

where $c > 0$ depends only on J.

Corollary 1.4

For every $J \geq 1$ there exist $f \in C_{\mathbb{R}^J}$ so that

(1.19) $\|\hat{f}\|_p = \infty$ for all $p < 2J/(J+1)$.

Proof

It suffices to show that for every $p < 2J/(J+1)$ and arbitrarily large $M > 0$ there exist $f \varepsilon C_{R^J}$ with a finitely supported transform so that $\|f\|_\infty = 1$ and $\|\hat{f}\|_p > M$. Let $N > 1$ be an arbitrary integer. By Lemma 1.3, there is a choice of signs so that

$$\| \sum_{n_1,\ldots,n_J=1}^{N} \pm r_{n_1} \otimes \cdots \otimes r_{n_J} \|_\infty \leq c\ N^{(J+1)/2} .$$

Write

$$f_N = \frac{1}{c\ N^{(J+1)/2}} \sum_{n_1,\ldots,n_J=1}^{N} \pm r_{n_1} \otimes \cdots \otimes r_{n_J} ,$$

and observe that $\|\hat{f}_N\|_p$ is an unbounded function of N if $p < 2J/(J+1)$.

$\boxed{\times}$

Remark 1.5

Corollary 1.4 was established in [8] by a use of Riesz products in a general context of harmonic analysis. The general J-dimensional inequalities of Theorem 1.2 were obtained in [8] by following Littlewood's own argument in the case $J = 2$. The proof by induction given here, beginning with the trivial case $J = 1$, is conceptually different and yields better constants. The starting point for both proofs, however, is the fundamental Khintchin inequality expressed through Lemma 1.1 whose statement is known as the J-dimensional Littlewood inequality.

The estimation

$$\kappa_{2-(2/J)} \leq \kappa^{1/(J-1)} ,$$

substituted in Theorem 1.2, implies

$$\lambda_J \leq \kappa (\lambda_{J-1})^{(J-1)/J} ,\ J > 1 ,$$

which, following an elementary computation, yields

(1.20) $\lambda_J \leq \kappa^{(\frac{J+1}{2} - \frac{1}{J})} ,\ J > 1 .$

When $J > 8$, the estimate in (1.20) can be improved via Haagerup's results [7].[2] $\lambda_2 = \kappa = \sqrt{2}$ is the best constant; the value of the best λ_J, $J > 2$, is unknown.

---

[2] $\kappa_p \leq \kappa^{(2-p)/p}$ follows from an interpolation argument. In [7], Haagerup shows $\kappa_p = \kappa^{(2-p)/p}$ for $0 < p \leq p_0$, $p_0 \approx 1.847...,$ and computes $\kappa_p < \kappa^{(2-p)/p}$ for $p_0 < p < 2$.

2. COMBINATORIAL DIMENSION AND FRACTIONAL CARTESIAN PRODUCTS

a. A definition of combinatorial dimension

We start with an infinite set X that is devoid of any structure. Fix a positive integer J and let X^J denote the usual J-fold Cartesian product of X,

$$X^J = \{(x_1, \ldots, x_J): \quad x_i \, \varepsilon \, X, \quad i = 1, \ldots, J\}.$$

We take an arbitrary subset $F \subset X^J$ and proceed to define its 'dimension.' Before matters are formalized, I will indulge briefly in a bit of heuristics and describe the point of view taken here. X^J could be regarded as an 'outcome space' for the following 'experiment:' Make J 'independent' samplings from the set X.[3] In this case,

$$\text{'dimension' of} \quad X^J = J$$

is the number of degrees of freedom enjoyed by the J samplings. The subset $F \subset X^J$ could be viewed also as an 'outcome space' for some prescribed experiment resulting in J selections $x_1, \ldots, x_J \varepsilon X$, the 'interdependencies' between which are apparent via the constraint that $(x_1, \ldots, x_J)$ be in F. And so, the 'dimension' of F should be an appropriate measurement of the degrees of freedom available to the J samplings in this case. Such a gauge is suggested by the following elementary counting principle: For every positive integer s and s-subsets $A_1, \ldots, A_J \subset X$ (sets with s elements), the number of possible selections of J elements $x_1 \varepsilon A_1, \ldots, x_J \varepsilon A_J$ equals s^J. Extending this principle to $F \subset X^J$, we want the 'dimension' of F to be a number, say dimF, with this property:

> For every positive integer s and s-subsets $A_1, \ldots, A_J \subset X$, the
> number of possible selections of J elements $x_1 \varepsilon A_1, \ldots, x_J \varepsilon A_J$,
> *subject to the requirement* that the outcome $(x_1, \ldots, x_J)$ be
> in F, is no greater than (roughly) s^{dimF}. On the other hand,

---

[3] 'Independence' means roughly this: information about any one of the samplings implies no information about the others.

there are arbitrarily large s and s-subsets $A_1, \ldots, A_J \subset X$ so
that the number of selections from $A_1, \ldots, A_J$, subject to the
requirement that the outcome be in F, is at least (roughly)
s^{dimF} .

We shall now be precise. Let s be a positive integer and define

(2.1) $\Psi_F(s) = \max \{ |F \cap (A_1 \times \cdots \times A_J)| : A_i \subset X, \ |A_i| = s, \ i = 1, \ldots, J \}$

(as usual, $|\cdot|$ denotes cardinality). To place the definition of Ψ_F in
the context of the discussion above, observe simply this: let $Q_1, \ldots, Q_J$
denote (here and throughout the paper) the canonical projections from
X^J onto X defined by

$$Q_i(x_1, \ldots, x_J) = x_i, \ i = 1, \ldots, J .$$

Then, (2.1) can be rewritten as

(2.1)' $\Psi_F(s) = \displaystyle\max_{\substack{A_1, \ldots, A_J \subset X \\ |A_1| = \cdots = |A_J| = s}} |\{ x \varepsilon F: \ Q_1(x) \varepsilon A_1, \ldots, Q_J(x) \varepsilon A_J \}|$

which is a measurement of interdependencies between the projections (the
J samplings) $Q_1, \ldots, Q_J$ restricted to F. Guided by the 'fractional'
counting principle described above, we define for every a > 0

(2.2) $d_F(a) = \sup_s \dfrac{\Psi_F(s)}{s^a}$,

and formalize

Definition 2.1

The *combinatorial dimension* of $F \subset X^J$ is

(2.3) $dimF = \inf\{a: d_F(a) < \infty \} .$

dimF is said to be *exact* if $d_F(dimF) < \infty$, and *asymptotic* if
$d_F(dimF) = \infty$.[4]

---

[4] In previous work [4], [5] we wrote $dimF = \inf\{a: \varlimsup_{s \to \infty}(\Psi_F(s)/s^a) < \infty\}$
which is equivalent to the present definition.

Remarks 2.2

(i) $d_F(\cdot)$ is a monotonically decreasing function satisfying

$$d_F(J) < \infty \quad (F \subset X^J),$$

$$\lim_{a \to 1^-} d_F(a) = \infty \quad \text{when } F \text{ is an infinite set},$$

and

$$\lim_{a \to 0^+} d_F(a) = |F| \quad \text{when } F \text{ is finite}.$$

Therefore, $1 \le \dim F \le J$ when $F \subset X^J$ is infinite, while the combinatorial dimension of finite sets equals 0.

To fix ideas, observe that if f is any function from X^2 into X
then

$$\dim\{(x,y,f(x,y)): (x,y) \varepsilon X^2\} = 2 \quad \text{exactly}.$$

Examples of sets with arbitrarily prescribed dimensions were produced
'randomly' in [5]. Explicit designs of fractionally dimensioned sets
(generalizations of examples in [3]) will be displayed in the next sub-
section.

(ii)   The following basic properties are easy to establish:

(a)   Let $F_1, F_2 \subset X^J$.   Then,

(2.4) $\dim(F_1 \cup F_2) = \max\{\dim F_1, \dim F_2\}$.

To verify (2.4), note that

$$\min\{\Psi_{F_1}(s), \Psi_{F_2}(s)\} \le \Psi_{F_1 \cup F_2}(s) \le 2 \max\{\Psi_{F_1}(s), \Psi_{F_2}(s)\}$$

for all s.

(b)   Let $F_1 \subset X^{J_1}$ and $F_2 \subset X^{J_2}$.   Then

(2.5) $\dim(F_1 \times F_2) \le \dim F_1 + \dim F_2$.

To verify (2.5), note that

$$\Psi_{F_1 \times F_2}(s) = \Psi_{F_1}(s) \cdot \Psi_{F_2}(s) \quad \text{for all s}.$$

b. Fractional Cartesian products of X

In what follows (here and throughout the paper), $J \geq K > 0$ denote fixed arbitrary integers, and

$$S_1, \ldots, S_N \subset \{1, \ldots, J\}$$

are subsets that satisfy the following:

(2.6) $|S_\alpha| = K$, $\alpha = 1, \ldots, N$;

(2.7)   there is a positive integer I so that for each $j \in \{1, \ldots, J\}$

$$|\{\alpha: j \in S_\alpha\}| = I$$

(I is the *incidence number* of $(S_\alpha)_{\alpha=1}^N$). Observe that

(2.8) $I = NK/J$.

Enumerate

$$S_\alpha = (\alpha_1, \ldots, \alpha_K), \quad 1 \leq \alpha_1 < \alpha_2 \cdots < \alpha_K \leq J,$$

and define projections

$$P_\alpha : X^J \longrightarrow X^K$$

by

(2.9) $P_\alpha(x_1, \ldots, x_J) = (x_{\alpha_1}, \ldots, x_{\alpha_K})$, $\alpha = 1, \ldots, N$.

We now fix a one-one correspondence between X and X^K designated by

(2.10) $(t_1, \ldots, t_K) \longleftrightarrow x(t_1, \ldots, t_K) \in X$, $t_1, \ldots, t_K \in X$,

(recall that X is infinite) and formalize

Definition 2.3

A J/K-fold Cartesian product of X is

(2.11) $X^{(J,K;N)} = \{(x(P_1(t)), \ldots, x(P_N(t))) : t \in X^J\} \subset X^N$.

Two extremal cases are singled out:

(i)   If $(S_\alpha)^N_{\alpha=1}$ is the collection of all K-subsets of

$\{1,\ldots,J\}$ (i.e., $N = \binom{J}{K}$ and $I = \binom{J-1}{K-1}$)   then

(2.12)
$$X^{(J,K;(\binom{J}{K}))} \equiv X^{J/K}$$

is called a maximal J/K-Cartesian product of X.

(ii)   If the incidence number of $(S_\alpha)^N_{\alpha=1}$ is K (i.e. N=J)   then

(2.13)
$$X^{(J,K;J)} \equiv X^{(J,K)}$$

is called a minimal J/K-Cartesian product of X.

The basic idea underlying the fractional Cartesian products is simply this: Points in X^J, formally given as J-tuples of *independent* coordinates can be described also as N-tuples of *interdependent* coordinates

$$x = (x_1,\ldots,x_J) <\longrightarrow (P_1(x),\ldots,P_N(x)), \quad x \varepsilon X^J,$$

which are then viewed as points of a *fractional* Cartesian product. The appropriate measurements of the interdependencies between $P_1,\ldots,P_N$ are based on the following general measure-theoretic estimate (a generalization of Lemma 2.2 in [3]):

Lemma 2.4

Suppose (X,dx) is measure space (with dx a positive measure). Let $f_1,\ldots,f_N$ be measurable functions on the product space X^K. Then

(2.14)
$$\int_{x \varepsilon X^J} |f_1(P_1(x)) \cdots f_N(P_N(x))| dx_1 \cdots dx_J \le \|f_1\|_I \cdots \|f_N\|_I$$

(the integral on the left hand side of (2.14) is a J-fold iterated integral over (X,dx); I is the incidence number of $(S_\alpha)^N_{\alpha=1}$; as usual, $\|f\|_I$ is the L^p-norm of f with p = I).

Proof

The inequality in (2.14) is obtained by J successive applications of the multilinear Hölder inequality

(2.15) $\int\limits_{X} |g_1(x) \cdots g_L(x)| dx \leq \|g_1\|_{p_1} \cdots \|g_L\|_{p_L}$,

where L is a positive integer, $g_1, \ldots, g_L$ are measurable functions on X,
and $1/p_1 + \cdots + 1/p_L = 1$.

Step 1: Assume (without loss of generality) that $1 \varepsilon S_1, \ldots, 1 \varepsilon S_I$.
Apply (2.15) with $L = I$ and $p_1 = \cdots = p_I = I$ to the integral over
$x_1 \varepsilon X$ and the functions $f_1 \circ P_1, \ldots, f_I \circ P_I$ on the left hand side of (2.14)
which thus becomes majorized by

$$\int\limits_{(x_2, \ldots, x_J) \varepsilon X^{J-1}} (\int\limits_{x_1} |f_1(P_1(x_1, \ldots, x_J))|^I dx_1)^{1/I} \cdots$$

$$\cdots (\int\limits_{x_1} |f_I(P_I(x_1, \ldots, x_J))|^I dx)^{1/I} .$$

$$\cdot |f_{I+1}(P_{I+1}(x_1, \ldots, x_J)) \cdots f_N(P_N(x_1, \ldots, x_J))| dx_2 \cdots dx_J .$$

The second step is similar: In the line above, again apply (2.15) with
$L = I$ and $p_1 = \cdots = p_I = I$ to the integral over $x_2 \varepsilon X$ and the I
factors indexed by $\{\alpha : 2 \varepsilon S_\alpha\}$ (recall property (2.7)). And so, following
successively J such steps based on (2.15) and (2.7), we obtain (2.14).

$\boxed{\text{x}}$

We apply Lemma 2.4 to the 'discrete' X of this section:

Theorem 2.5

(i) Let $A_1, \ldots, A_N$ be arbitrary finite subsets of X. Then

(2.16) $|X^{J, K; N)} \cap (A_1 \times \cdots \times A_N)| \leq (|A_1| \cdots |A_N|)^{1/I}$.

In particular,

(2.17) $\Psi_{X}(J, K; N)(s) \leq s^{J/K}$ for all positive integers s.

(ii) For every integer $s > 0$ there is $A \subset X$, $|A| = s^K$, so that

(2.18) $|X^{(J, K; N)} \cap A^N| = |A|^{J/K}$.

In particular,

(2.19) $\Psi_X(J,K;N)(s) \geq Cs^{J/K}$ for all integers $s > 0$

where $C > 0$ depends only on   J   and   K.

<u>Proof</u>

(i)   Denote by χ_α the indicator function of A_α, $\alpha = 1,\ldots,N$:

$$\chi_\alpha(x) = \begin{cases} 1 & \text{if } x \in A_\alpha \\ 0 & \text{otherwise} . \end{cases}$$

Given the definition of $X^{(J,K;N)} \subset X^N$,   observe that

(2.20) $|X^{(J,K;N)} \cap (A_1 \times \cdots \times A_N)| = \sum_{x \in X^J} \chi_1(P_1(x)) \cdots \chi_N(P_N(x))$.

Now apply Lemma 2.4 to right hand side of (2.20) with $f_\alpha = \chi_\alpha$,
$\alpha = 1,\ldots,N$,   on the 'discrete' measure space   X   with dx = 'counting'
measure.   Note that $\|\chi_\alpha\|_I = |A_\alpha|^{1/I}$ and thus deduce (2.16).

To deduce (2.17), merely apply the definition of Ψ (in (2.1)) and
property (2.8), $N/I = J/K$.

(ii)   Fix any $B \subset X$ whose cardinality is   s.   Let (following (2.10))

$$A = \{x(t_1,\ldots,t_K) : t_1,\ldots,t_K \in B\} \subset X .$$

Denote the indicator function of   A   by χ_A and compute

(2.21) $|X^{(J,K;N)} \cap A^N| = \sum_{t \in X^J} \chi_A(x(P_1(t))) \cdots \chi_A(x(P_N(t)))$

$$= |A|^{J/K} = s^J ,$$

establishing (2.18).

(2.21) implies

$$\Psi_X(J,K;N)(s^K) = s^J \quad \text{for all integers} \quad s > 0 .$$

Therefore, given any $s > 0$,   let   n   be the integer between $s^{1/K} - 1$ and
$s^{1/K}$ and estimate

$$\Psi_X(J,K;N)^{(s)} \geq \Psi_X(J,K;N)^{(n^K)}$$

$$= n^J \geq (s^{1/K} - 1)^J ,$$

establishing (2.19). $\boxed{x}$

Combining (2.17) and (2.19), we obtain

Corollary 2.6

(2.22) $$1 \geq d_X(J,K;N)^{(J/K)} > \inf_s \frac{\Psi_X(J,K;N)^{(s)}}{s^{J/K}} \geq C > 0 .$$

In particular,

(2.23) $$\dim X^{(J,K;N)} = J/K \quad \text{exactly} .$$

Remark 2.7

Property (2.23) is weaker than property (2.22) which we formalize:
$F \subset X^J$ is an α-Cartesian product if

$$\infty > d_F(\alpha) \geq \inf_s \frac{\Psi_F(s)}{s^\alpha} > 0 .$$

Problem: Given an arbitrary $1 < \alpha < 2$, produce an α-Cartesian product
in $\mathbb{N}^2$.

c. Measure-theoretic isoperimetric inequalities

The computation of dimensions of fractional Cartesian products is
closely related to so called isoperimetric inequalities (for a general
reference, see [16] and, in particular, section 2 thereof). Suppose
(X,dx) is a measure space. Let D be a measurable subset in the product
measure space X^J, and denote 'volume' of $D = v_J(D) = \int_D dx_1 \cdots dx_J$.
Let $(S_\alpha)_{\alpha=1}^J$ be a (minimal) collection of K-subsets of $\{1,\ldots,J\}$
(incidence number $= K$), and $P_1,\ldots,P_J$ be the projections from X^J onto
X^K defined by (2.9). Denote the image of D under P_α by $P_\alpha[D]$, and
the inverse image of a subset $E \subset X^K$ by $P_\alpha^{-1}[E]$, $\alpha = 1,\ldots,J$. We clearly
have

$$D \subset \bigcap_{\alpha=1}^{J} P_{\alpha}^{-1}[P_{\alpha}[D]],$$

and therefore

$$\nu_J(D) \leq \nu_J(\bigcap_{\alpha=1}^{J} P_{\alpha}^{-1}[P_{\alpha}[D]])$$

(2.24)
$$= \int_{X^J} \chi_1(P_1(x)) \cdots \chi_J(P_J(x)) dx_1 \cdots dx_J,$$

where χ_{α} denotes the indicator function of $P_{\alpha}[D]$ on X^K, $\alpha = 1, \ldots, J$. Applying Lemma 2.4 to the right hand side of (2.24), we obtain

Theorem 2.8

(2.25)
$$\nu_J(D) \leq \left(\prod_{j=1}^{J} \nu_K(P_j[D]) \right)^{1/K}.$$

By the usual 'geometric-arithmetic mean' inequality, we deduce from (2.25) that

$$(\nu_J(D))^{K/J} \leq \frac{1}{J} \sum_{j=1}^{J} \nu_K(P_j[D]),$$

which could be viewed as a measure-theoretic isoperimetric inequality.

Professor Amram Meir kindly pointed out to us that Loomis and Whitney [13] had used an inequality similar to (2.25) (with a different proof) in Euclidean space and the case $K = J - 1$ to deduce a 'qualitative' version of the classical isoperimetric inequality (inequality (2.7) in [16]).

3. COMBINATORIAL DIMENSION AND BOUNDED FRACTIONAL FORMS

We preserve notation of the previous two sections: Throughout, $J \geq K > 0$ denote arbitrary fixed integers, and $(S_\alpha)_{\alpha=1}^N$ denotes a collection of K-subsets of $\{1,\ldots,J\}$ satisfying (2.7). We take X (of section 2) to be R, the system of Rademacher functions (of section 1) indexed by $\mathbb{N}^K$,

(3.1)
$$R = \{r_k\}_{k \in \mathbb{N}^K} .$$

Such an enumeration of R (step (2.10) in section 2) is expressed here through a correspondence between $\mathbb{N}^K$ and $\mathbb{N}$,

(3.2)
$$\mathbb{N}^K \ni (k_1,\ldots,k_K) \longleftrightarrow n(k_1,\ldots,k_K) \in \mathbb{N} ,$$

by writing

$$R = \{r_{n(k)}\}_{k \in \mathbb{N}^K} ,$$

where $R = \{r_n\}_{n \in \mathbb{N}}$ is the usual enumeration. As in section 2, define

$$P_\alpha(j_1,\ldots,j_J) = (j_{\alpha_1},\ldots,j_{\alpha_K})$$

where $S_\alpha = (\alpha_1,\ldots,\alpha_K)$, $\alpha = 1,\ldots,N$. Now write a J/K-Cartesian product

$$R^{(J,K;N)} = \{r_{P_1(j)} \otimes \cdots \otimes r_{P_N(j)} : j \in \mathbb{N}^J\}$$

which we identify with the J/K-Cartesian product

$$\mathbb{N}^{(J,K;N)} = \{(n(P_1(j)),\ldots,n(P_N(j))) : j \in \mathbb{N}^J\} .$$

A subset $F \subset R^{(J,K;N)}$ is identified with

(3.3)
$$\{(n_1,\ldots,n_N) \in \mathbb{N}^N : r_{n_1} \otimes \cdots \otimes r_{n_N} \in F\} \subset \mathbb{N}^{(J,K;N)}$$

which will be designated also F. Observe that $F \subset \mathbb{N}^{(J,K;N)}$ is uniquely determined by

$$\Lambda_F = \{j \varepsilon \mathbb{N}^J : \big(n(P_1(j)), \ldots, n(P_N(j))\big) \varepsilon F\} \ .$$

In reverse, given $\Lambda \subset \mathbb{N}^J$, we denote the corresponding subset of $\mathbb{N}^{(J,K;N)}$ by

(3.4) $$F_\Lambda = \{\big(n(P_1(j)), \ldots, n(P_N(j))\big) : j \varepsilon \Lambda\} \ .$$

Extending the notion of multilinear forms given by (1.9), we shall deal with scalar valued functions on $[0,1]^N$, assumed integrable, that are spanned by $R^{(J,K;N)}$. These are N-linear forms f uniquely determined (via (1.9)) by J-tensors:

(3.5) $$\begin{cases} \hat{f}\big(n(P_1(j)), \ldots, n(P_N(j))\big) = a_j \ , & j \varepsilon \mathbb{N}^J \\[2em] \hat{f}(n_1, \ldots, n_N) \qquad\qquad = 0 \ , & (n_1, \ldots, n_N) \notin \mathbb{N}^{(J,K;N)} \ . \end{cases}$$

The space of all such forms in $C_{R^N}([0,1]^N)$ is denoted $C_{R}(J,K;N)$; its elements will be called bounded J/K-linear forms. Given $f \varepsilon C_{R}(J,K;N)$ and $F \subset \mathbb{N}^{(J,K;N)}$, denote the restriction of $\hat{f}$ to F by $\hat{f}\big|_F$ and its ℓ^P-norm

$$\|\hat{f}\big|_F\|_p = \Big(\sum_{(n_1, \ldots, n_N) \varepsilon F} |\hat{f}(n_1, \ldots, n_N)|^P\Big)^{1/p} \ .$$

The space of bounded J/K-linear forms whose transform is supported in $F \subset \mathbb{N}^{(J,K;N)}$ is denoted

$$C_F = \{f \varepsilon C_{R}(J,K;N) : \hat{f}(n) = 0, \quad n \notin F\} \ .$$

The setting for the theorem below, the main result of this section is the maximal J/K-Cartesian product of R which is denoted by $R^{J/K}$ (Definition 2.3(i)).

Theorem 3.1

Let Λ be an infinite subset of $\mathbb{N}^J$, and define

(3.6) $$p(\Lambda, K) = p = \max\{1, 2/(1+K/\dim\Lambda)\} \ .$$

(i) Suppose $\dim\Lambda$ is exact. Then:

(3.7) $\qquad \|\hat{f}|_{F_\Lambda}\|_p \le \beta\|f\|_\infty$ for all $f\epsilon C_{R^{J/K}}$,

where $\beta > 0$ depends on J,K and dimΛ .

(3.7)$^{\#}$ $\qquad\qquad$ There exist $f\epsilon C_{F_\Lambda}$ so that

$$\|\hat{f}\|_q = \|\hat{f}|_{F_\Lambda}\|_q = \infty \quad \text{for all}\quad q < p \; .$$

$\qquad$ (ii) Suppose dimΛ is asymptotic. Then:

(3.8) $\qquad\qquad \|\hat{f}|_{F_\Lambda}\|_r \le \beta_r\|f\|_\infty$ for all $f\epsilon C_{R^{J/K}}$

and all $r > p$, where $\beta_r > 0$ depends on r,J,K and dimΛ .

(3.8)$^{\#}$ $\qquad\qquad$ There exist $f\epsilon C_{F_\Lambda}$ so that

$$\|\hat{f}\|_p = \infty \; .$$

(β and β_r denote the 'best' constants in the inequalities (3.7) and
(3.8); F_Λ is defined by (3.4) with $N = \binom{J}{K}$.)

$\qquad$ In what follows below, $\sum_{S_\alpha}$ and $\sum_{S_{\tilde{\alpha}}}$ denote summation with respect
to those variables among $j_1,\dots,j_J$ which are indexed by $S_\alpha = (\alpha_1,\dots,\alpha_K)$
and $\{1,\dots,J\} \sim S_\alpha$, respectively.

Lemma 3.2

$\qquad$ Suppose $f\epsilon C_R(J,K;N)$,

$$f \sim \sum_{j\epsilon \mathbb{N}^J} a_j r_{P_1}(j) \otimes \cdots \otimes r_{P_N}(j) \; .$$

Then

(3.9) $\qquad\qquad \kappa^{[J/K]}\|f\|_\infty \ge \sum_{S_\alpha} \left(\sum_{S_{\tilde{\alpha}}} |a_j|^2\right)^{1/2}, \quad \alpha = 1,\dots,N$

($\kappa = \sqrt{2}$ is the Khintchin constant defined in (1.7); [J/K] is the
greatest integer less than J/K).

Proof

Without loss of generality take $\alpha = 1$.

$$\|f\|_\infty = \sup_{t_1,\ldots,t_N} |\sum_{j \in \mathbb{N}^J} a_j r_{P_1(j)}(t_1) \cdots r_{P_N(j)}(t_N)|$$

$$= \sup_{t_2,\ldots,t_N} (\sup_{t_1} |\sum_{S_1} (\sum_{\tilde{S}_1} a_j r_{P_2(j)}(t_2) \cdots r_{P_N(j)}(t_N)) r_{P_1(j)}(t_1)|)$$

$$= \sup_{t_2,\ldots,t_N} \sum_{S_1} | \sum_{\tilde{S}_1} a_j r_{P_2(j)}(t_2) \cdots r_{P_N(j)}(t_N)| \qquad \text{(by (1.6))}$$

$$(3.10) \qquad \geq \sum_{S_1} \int_{[0,1]^{N-1}} |\sum_{\tilde{S}_1} a_j r_{P_2(j)}(t_2) \cdots r_{P_N(j)}(t_N)| \, dt_2 \cdots dt_N \, .$$

We claim the following:

$$\kappa^{[J/K]} \int_{[0,1]^{N-1}} |\sum_{\tilde{S}_1} a_j r_{P_2(j)}(t_2) \cdots r_{P_N(j)}(t_N)| \, dt_2 \cdots dt_N \geq (\sum_{\tilde{S}_1} |a_j|^2)^{1/2} \, .$$

Observe that $[J/K]$ subsets among $S_2,\ldots,S_N$, say $S_{\alpha(1)},\ldots,S_{\alpha([J/K])}$, satisfy

$$\bigcup_{\ell=1}^{[J/K]} S_{\alpha(\ell)} = \{K+1,\ldots,J\} \, .$$

Now integrate $|\sum_{\tilde{S}_1} a_j r_{P_2(j)} \otimes \cdots \otimes r_{P_N(j)}|$ over $[0,1]^{[J/K]}$ with respect to $t_{\alpha(1)},\ldots,t_{\alpha([J/K])}$ and apply the multilinear Khintchin inequality (1.10) to establish the claim.

(3.9) follows from (3.10) and the claim.   ☒

Lemma 3.3

Suppose a scalar valued J-tensor $(\phi(j))_{j \in \mathbb{N}^J}$ satisfies the following:

(3.11) For every positive integer L and subsets $A_1,\ldots,A_J \subset \mathbb{N}$,

$$|A_1| = \cdots = |A_J| = L \, ,$$

$$\sum_{j \in A_1 \times \cdots \times A_J} |\phi(j)| \leq L^K \, .$$

Then:

(3.12)   For every positive integer L and subsets $A_1,\ldots,A_J \subset \mathbb{N}$,

$|A_1| = \cdots = |A_J| = L$, there exists a partition $\{T_1,\ldots,T_{\binom{J}{K}}\}$

of $A_1 \times \cdots \times A_J$ so that

$$\sup_{S_\alpha} \sum_{\tilde{S}_\alpha} |\phi(j)| \chi_{T_\alpha}(j) \le \binom{J}{K}$$

for each $\alpha = 1,\ldots,\binom{J}{K}$.

(Notation: As usual, $S_1,\ldots,S_{\binom{J}{K}}$ denote all K-subsets of $\{1,\ldots,J\}$;

$\sup_{S_\alpha}$ denotes supremum over $j_{\alpha_1},\ldots,j_{\alpha_K}$; χ_{T_α} is the indicator function

of T_α .)

Proof

Let $A_1,\ldots,A_J \subset \mathbb{N}$ and fix $j \varepsilon A_1 \times \cdots \times A_J$. For each $\alpha = 1,\ldots,\binom{J}{K}$,
define

(3.13) $\mathrm{Plane}(j)_\alpha = P_\alpha^{-1}[P_\alpha(j)] \cap (A_1 \times \cdots \times A_J)$,

and

$$\phi_\alpha(j) = \sum_{i \varepsilon \mathrm{Plane}(j)_\alpha} |\phi(i)|$$

($\mathrm{Plane}(j)_\alpha$ is the (J-K)-dimensional hyperplane in $A_1 \times \cdots \times A_J$ which
passes through j and is 'perpendicular' to the K-dimensional coordinate
plane $P_\alpha[\mathbb{N}^J]$) .

Sublemma

Suppose $|A_1| = \cdots = |A_J| = L > 0$. Then, for some $j \varepsilon A_1 \times \cdots \times A_J$

(3.14) $$\sum_{\alpha=1}^{\binom{J}{K}} \phi_\alpha(j) \le \binom{J}{K}$$.

<u>Proof</u>

To obtain (3.14) for some $j \varepsilon A_1 \times \cdots \times A_J$ with $|A_1 \times \cdots \times A_J| = L^J$, we show that

$$(3.15) \qquad \sum_{j \varepsilon A_1 \times \cdots \times A_J} \sum_{\alpha=1}^{\binom{J}{K}} \phi_\alpha(j) \leq L^J \binom{J}{K} .$$

Fix $1 \leq \alpha \leq \binom{J}{K}$. Under the hypothesis, each hyperplane given by (3.13) contains L^{J-K} points. Moreover, if $i \varepsilon \text{Plane}(j)_\alpha$ then

$$\text{Plane}(j)_\alpha = \text{Plane}(i)_\alpha .$$

And so, each point in $A_1 \times \cdots \times A_J$ appears in the enumeration

$$\{\text{Plane}(j)_\alpha : j \varepsilon A_1 \times \cdots \times A_J\}$$

L^{J-K} times. Therefore, we have

$$\sum_{j \varepsilon A_1 \times \cdots \times A_J} \sum_{\alpha=1}^{\binom{J}{K}} \phi_\alpha(j) = \sum_{\alpha=1}^{\binom{J}{K}} \sum_{j \varepsilon A_1 \times \cdots \times A_J} \sum_{i \varepsilon \text{Plane}(j)_\alpha} |\phi(i)|$$

$$= L^{J-K} \sum_{\alpha=1}^{\binom{J}{K}} \sum_{j \varepsilon A_1 \times \cdots \times A_J} |\phi(j)| .$$

Applying the assumption (3.11) to the line above, we obtain (3.15).   ☒

Lemma 3.3 is proved by induction on L: The case $L = 1$ is trivial. Let $L > 1$ and assume (3.12) holds in the case $L - 1$. Fix arbitrary subsets $A_1, \ldots, A_J \subset \mathbb{N}$, $|A_1| = \cdots |A_J| = L$. By the Sublemma, there exists

$$j^o = (j_1^o, \ldots, j_J^o) \varepsilon A_1 \times \cdots \times A_J$$

so that (a fortiori) for each $\alpha = 1, \ldots, \binom{J}{K}$

$$(3.16) \qquad \sum_{i \varepsilon \text{Plane}(j^o)_\alpha} |\phi(i)| \leq \binom{J}{K} .$$

Now delete j_ℓ^o from A_ℓ:

$$A_\ell' = A_\ell \sim \{j_\ell^o\} , \quad \ell = 1, \ldots, J .$$

Apply the induction hypothesis to $A_1', \ldots, A_J'$ and obtain a partition $\{T_1', \ldots, T_{\binom{J}{K}}'\}$ of $A_1' \times \cdots \times A_J'$ for which (3.12) holds. Define

(3.17) $$T_\alpha = T_\alpha' \cup \text{Plane}(j^0)_\alpha, \qquad \alpha = 1, \ldots \binom{J}{K}.$$

The induction hypothesis and (3.16) imply that $\{T_1, \ldots, T_{\binom{J}{K}}\}$ is the required partition in (3.12).   ☒

Lemma 3.5

Given $\Lambda \subset \mathbb{N}^J$, suppose

(3.18) $$d_\Lambda(b) \leq c \qquad (d_\Lambda \text{ is defined by (2.2)}).$$

Let ϕ be a J-tensor with a finite support contained in Λ. Suppose that ϕ satisfies

(3.19)
$$
\begin{cases}
\sum_{j \in \mathbb{N}^J} |\phi(j)|^{2/(1-K/b)} \leq 1 & \text{if } b > K \\[2ex]
\sup_{j \in \mathbb{N}^J} |\phi(j)| \leq 1 & \text{if } 0 < b \leq K.
\end{cases}
$$

Then: For all $f \in C_{\mathbb{R}^{J/K}}$,

$$f \sim \sum_{j \in \mathbb{N}^J} a_j r_{P_1}(j) \otimes \cdots \otimes r_{P_N}(j),$$

we have

(3.20) $$\sum_{j \in \mathbb{N}^J} |a_j \phi(j)| \leq c_b \binom{J}{K}^{3/2} \kappa^{[J/K]} \|f\|_\infty$$

where $c_b > 0$ depends on c and b (of (3.18)).

Proof

First, we verify that $|\phi|^2$ satisfies hypothesis (3.11) of Lemma 3.3:

Let $A_1, \ldots, A_J$ be arbitrary subsets of $\mathbb{N}$, $|A_1| = \cdots = |A_J| = L > 1$. Apply Hölder's inequality to

$$\sum_{j \varepsilon A_1 \times \cdots \times A_J} |\phi(j)|^2 = \sum_{j \in \mathbb{N}^J} |\phi(j)|^2 \chi_{\Lambda \cap (A_1 \times \cdots \times A_J)} (j)$$

with exponents

$$\left\{ \begin{array}{ccc} b/(b-K) & \text{and } b/K & \text{if } b > K \\ \infty & \text{and } 1 & \text{if } b \leq K \end{array} \right.$$

to $|\phi|^2$ and $\chi_{\Lambda \cap (A_1 \times \cdots \times A_J)}$, respectively. We thus obtain

(3.21)
$$\sum_{j \varepsilon A_1 \times \cdots \times A_J} |\phi(j)|^2 \leq \left\{ \begin{array}{ll} |\Lambda \cap (A_1 \times \cdots \times A_J)|^{K/b} \sum_{j \in \mathbb{N}^J} |\phi(j)|^{2/(1-K/b)} , & b > K \\ \\ |\Lambda \cap (A_1 \times \cdots \times A_J)| \sup_j |\phi(j)|^2 , & b \leq K . \end{array} \right.$$

But, the assumption (3.18) means:

$$\Psi_\Lambda(L) \leq cL^b \quad \text{for all } L \geq 1 ,$$

and, in particular, $|\Lambda \cap (A_1 \times \cdots \times A_J)| \leq cL^b$. Therefore, (3.21) and (3.19) combined imply

$$\sum_{j \varepsilon A_1 \times \cdots \times A_J} |\phi(j)|^2 \leq \left\{ \begin{array}{ll} c^{K/b} L^K , & b > K \\ \\ cL^b , & b \leq K , \end{array} \right.$$

which establishes that $|\phi|^2$ satisfies (3.11) 'scaled' by c; for typographical reasons, we assume now (without loss of generality) that c = 1.

Suppose that the support of ϕ is contained in some fixed 'cube' $A_1 \times \cdots \times A_J$, $|A_1| = \cdots = |A_J| = L$. Having verified (3.11) for $|\phi|^2$, we produce a partition $\{T_1, \ldots, T_{\binom{J}{K}}\}$ of $A_1 \times \cdots \times A_J$ so that for every $\alpha = 1, \ldots, \binom{J}{K}$ we have

(3.22)
$$\sup_{S_\alpha} (\sum_{\tilde{S}_\alpha} |\phi(j)|^2 \chi_{T_\alpha}(j))^{1/2} \leq \binom{J}{K}^{1/2} .$$

Estimate

$$\sum_{j \in \mathbb{N}^J} |a_j \phi(j)| \leq \sum_{\alpha=1}^{\binom{J}{K}} \sum_{j \in \mathbb{N}^J} |a_j \phi(j) \chi_{T_\alpha}(j)|$$

$$\leq \sum_{\alpha=1}^{\binom{J}{K}} \sup_{S_\alpha} (\sum_{\tilde{S}_\alpha} |\phi(j)|^2 \chi_{T_\alpha}(j))^{1/2} \sum_{S_\alpha} \sum_{\tilde{S}_\alpha} |a_j|^2)^{1/2}$$

(by Hölder's inequality applied to $\sum\limits_{S_\alpha}$)

$$\leq \binom{J}{K}^{3/2} K^{[J/K]} \|f\|_\infty$$

(by (3.22) and (3.9) of Lemma 3.2),

which establishes (3.20).   ⊠

Proof of Theorem 3.1

The inequalities (3.7) and (3.8) follow by the elementary duality $(\ell^p)^* = \ell^q$, $\frac{1}{p} + \frac{1}{q} = 1$, from Lemma 3.5, and the definition

$$\dim \Lambda = \inf\{b : d_\Lambda(b) < \infty\}$$

which is exact if $d_\Lambda(\dim \Lambda) < \infty$, and asymptotic if $d_\Lambda(\dim \Lambda) = \infty$.

To prove (3.7)[#] and (3.8)[#], again in view of the definition of $\dim \Lambda$, it suffices to establish the following

Lemma 3.6

Suppose $b \geq K$ and

(3.23) $d_\Lambda(b) = \infty$.

Then: For every $M > 0$ there exist $f \in C_{F_\Lambda}$,

$$f \sim \sum_{j \in \Lambda} a_j r_{P_1(j)} \otimes \cdots \otimes r_{P_{\binom{J}{K}}(j)}$$

so that $\{j \in \Lambda : a_j \neq 0\}$ is finite, $\|f\|_\infty \leq 1$ and

$$\sum_{j \in \Lambda} |a_j|^{2/(1+K/b)} > M.$$

Proof

Fix a large $D > 0$. By assumption (3.23), for arbitrarily large s there exist $A_1, \ldots, A_J \subset \mathbb{N}$, $|A_1| = \cdots = |A_J| = s$, so that

(3.24)
$$|\Lambda \cap (A_1 \times \cdots \times A_J)| > Ds^b .$$

Write

$$\Lambda_s = \Lambda \cap (A_1 \times \cdots \times A_J) .$$

Observe that (by (3.1) and (3.2))

$$\{ r_{P_1(j)} \otimes \cdots \otimes r_{P_{\binom{J}{K}}(j)} : j \varepsilon A_1 \times \cdots \times A_J \} \subset B_1 \times \cdots \times B_{\binom{J}{K}}$$

where

$$B_\alpha = \{ r_k \}_{k \varepsilon A_{\alpha_1} \times \cdots \times A_{\alpha_K}}$$

and $|B_\alpha| = s^K$ for each $\alpha = 1, \ldots, \binom{J}{K}$. Therefore, by the Kahane-Salem-Zygmund probabilistic estimates, via Lemma 1.3, we have a choice of signs $\pm$ so that

$$\| \sum_{j \varepsilon \mathbb{N}^J} \pm \chi_{\Lambda_s}(j) r_{P_1(j)} \otimes \cdots \otimes r_{P_{\binom{J}{K}}(j)} \|_\infty \leq c(|\Lambda_s| s^K)^{1/2} ,$$

where $c > 0$ is a constant depending only on $\binom{J}{K}$. Define

$$f_s = \left(\sum_{j \varepsilon \Lambda_s} \pm r_{P_1(j)} \otimes \cdots \otimes r_{P_{\binom{J}{K}}(j)} \right) / c(|\Lambda_s| s^K)^{1/2} ,$$

and estimate by (3.24)

$$\| \hat{f}_s \|_{2/(1+K/b)}^{2/(1+K/b)} \leq D^{1/2} s^{(b-K)/2} / c .$$

$D > 0$ is arbitrary, and so f_s is the required function. $\boxed{\times}$

Remark 3.7

The constants β and β_r in the inequalities (3.7) and (3.8) were majorized here, via (3.20) in the statement of Lemma 3.5, by constant multiples of $\binom{J}{K}^{3/2} K^{[J/K]}$. In the case $\Lambda = \mathbb{N}^J$, this estimation can

be replaced by the computation in [3] based on a general 'pure'
inequality:

Lemma (Lemma 5.3 in [3])

Let (X,dx) be a measure space and f be a measurable function on
the product space X^J. Then

$$\left(\int_{X^J} |f(x_1,\ldots,x_J)|^{2/(1+K/J)} dx_1 \cdots dx_J\right)^{(1+K/J)/2} \leq$$

$$\leq \prod_{\alpha=1}^{\binom{J}{K}} \left(\sum_{S_\alpha} \left(\int_{S_\alpha^\sim} |f(x_1,\ldots,x_J)|^2\right)^{1/2}\right)^{1/\binom{J}{K}}$$

($\int_{S_\alpha}$ and $\int_{S_\alpha^\sim}$ denote integration with respect to the variables indexed
by S_α and $\{1,\ldots,J\} \sim S_\alpha$, respectively).

Therefore, combining Lemma 3.2 and the Lemma above, we obtain in the case
$\Lambda = \mathbb{N}^J$ that $\beta \leq \kappa^{[J/K]}$. In any case, for an arbitrary $\Lambda \subset \mathbb{N}^J$, the
precise dependence of β on $\dim\Lambda$, $d_\Lambda(\cdot)$, J and K is not known; this
problem is more general and difficult than the one suggested in Remark 1.5.

The inequalities involving the J/K-Cartesian products $R^{(J,K;N)}$,
$J \leq N < \binom{J}{K}$, are typically obtained as an instance of Theorem 3.1:

Corollary 3.8

$$\|\hat{f}\|_p < \infty \quad \text{for all} \quad f \varepsilon C_{R(J,K;N)}$$

if and only if

$$p \geq 2/(1+K/J) .$$

Proof

Write $\Lambda = \mathbb{N}^{(J,K;N)} \subset \mathbb{N}^N$. By Corollary 2.6, $\dim\Lambda = J/K$ exactly.
Apply Theorem 3.1 with $J = N$ and $K = 1$.   ☒

Corollary 3.8 suggests an 'analytic' parameter associated with an arbitrary $F \subset R^N$:

$$(3.25) \qquad \sigma_F = \inf\{q: \sup_{\substack{f \in C_F \\ f \neq 0}} \|\hat{f}\|_q / \|f\|_\infty < \infty\} ;$$

σ_F is called the Sidon exponent of F and said to be

$$
\left\{
\begin{array}{lll}
\text{exact} & \text{if} & \displaystyle\sup_{0 \neq f \in C_F} \|\hat{f}\|_{\sigma_F} / \|f\|_\infty < \infty \\[2em]
\text{asymptotic if} & & \displaystyle\sup_{0 \neq f \in C_F} \|\hat{f}\|_{\sigma_F} / \|f\|_\infty = \infty
\end{array}
\right.
$$

(e.g., Definition 6.2 in [4]). Theorem 3.1 clearly implies the following

Corollary 3.9

Suppose $\Lambda \subset \mathbb{N}^J$ is infinite, and (as in (3.4)) define

$$F_\Lambda = \{r_{P_1(j)} \otimes \cdots \otimes r_{P_{\binom{J}{K}}(j)} : j \in \Lambda\} \subset R^{\binom{J}{K}} .$$

Then:

$$\sigma_{F_\Lambda} = \max\{1, 2/(1+\dim\Lambda/K)\} \quad (= p(\Lambda, K) \quad \text{in (3.6)}).$$

Moreover, σ_{F_Λ} is exact if and only if $\dim\Lambda$ is exact'.

Corollary 3.9 implies (and, in fact, is equivalent to) the following combinatorial fact:

Corollary 3.10

Suppose $\Lambda \subset \mathbb{N}^J$ is infinite, and

$$F_\Lambda = \{(n(P_1(j)), \ldots, n(P_{\binom{J}{K}}(j))) : j \in \Lambda\} .$$

Then:

(3.26) $\dim F_\Lambda = \max\{1, \dim \Lambda / K\}$,

where $\dim F_\Lambda$ is exact if and only if $\dim \Lambda$ is exact.

Proof

As usual, following (3.1) and (3.2), view F_Λ also as a subset of $R^{J/K} \subset R^{\binom{J}{K}}$. By Corollary 3.9,

$$\sigma_{F_\Lambda} = \max\{1, 2/(1+\dim \Lambda / K)\} .$$

But again applying Corollary 3.9 to $F_\Lambda \subset \mathbb{N}^{\binom{J}{K}}$, we deduce also

$$\sigma_{F_\Lambda} = 2/(1+\dim F_\Lambda) .$$

Combining the two equalities above, which are jointly exact or jointly asymptotic, we obtain the desired conclusion. $\boxed{x}$

Remark 3.11

Corollary 3.10 is, of course, completely general: We take any infinite set X (of section 2), any infinite subset $\Lambda \subset X^J$, define (following (2.10))

$$F_\Lambda = \{(x(P_1(t)), \ldots, x(P_{\binom{J}{K}}(t))) : t \varepsilon \Lambda\} \subset X^{\binom{J}{K}},$$

and deduce

(3.26)' $\dim F_\Lambda = \max\{1, \dim \Lambda / K\}$.

(3.26)' relates two types of measurements: On the one side, the combinatorial dimension of $\Lambda \subset X^J$ measures the interdependencies between the restrictions to Λ of $Q_1, \ldots, Q_J$, the canonical projections from X^J onto the respective J 'coordinate axes.' On the other side, the combinatorial dimension of F_Λ measures the interdependencies between the restrictions to Λ of $P_1, \ldots, P_{\binom{J}{K}}$, the canonical projections from X^J onto the respective $\binom{J}{K}$ 'mutually orthogonal K-dimensional coordinate planes' of X^J. In effect, $\dim F_\Lambda$ could be viewed as the combinatorial dimension of $\Lambda \subset X^J$ *relative* to the $\binom{J}{K}$ 'interdependent' K-dimensional

coordinate planes $P_1[X^J], \ldots, P_{\binom{J}{K}}[X^J]$.

Question: Let $(S_\alpha)_{\alpha=1}^N$ be a collection of K-subsets of $\{1, \ldots, J\}$

satisfying (2.7) with $J \le N < \binom{J}{K}$. Suppose $\Lambda \subset X^J$ is arbitrary.

What can be said about the combinatorial dimension of

$$F_\Lambda = \{(x(P_1(t)), \ldots, x(P_N(t))) : t\varepsilon\Lambda\} \ ?$$

4. FRÉCHET PSEUDOMEASURES

a. J-dimensional Fréchet Pseudomeasures

Let $(X_1, A_1), \ldots, (X_J, A_J)$ be measurable spaces and assume that each of the σ-algebras $A_1, \ldots, A_J$ is infinite. A measurable rectangle in the J-fold Cartesian product $\underset{j=1}{\overset{J}{\times}} X_j$ will be a set of the form $E_1 \times \cdots \times E_J$, $E_1 \in A_1, \ldots, E_J \in A_J$. As usual, $\underset{j=1}{\overset{J}{\times}} A_j$ will denote the product σ-algebra generated by the measurable rectangles and $(\underset{j=1}{\overset{J}{\times}} X_j, \underset{j=1}{\overset{J}{\times}} A_j)$ will designate the corresponding measurable product space. A partition of a measurable set E will mean here a countable collection of mutually disjoint measurable sets whose union is E.

Definition 4.1

A scalar valued function μ defined on the measurable rectangles in $\underset{j=1}{\overset{J}{\times}} X_j$ is an F_J-pseudomeasure if the following holds for every $E_1 \in A_1, \ldots, E_J \in A_J$:

(4.1) For every $j \in \{1, \ldots, J\}$ and every partition $\{F_j(n)\}_{n \in \mathbb{N}}$ of E_j

$$\sum_{n \in \mathbb{N}} \mu(E_1 \times \cdots \times F_j(n) \times \cdots \times E_J) = \mu(E_1 \times \cdots \times E_J) .$$

The space of F_J-pseudomeasures on $\underset{j=1}{\overset{J}{\times}} X_j$ is denoted by $F_J(\underset{j=1}{\overset{J}{\times}} x_j) = F_J$. (5

The requirement (4.1) means simply this: For every $E_1 \in A_1, \ldots, E_J \in A_J$ and each $j \in \{1, \ldots, J\}$ the function μ_j defined by

$$\mu_j(F) = \mu(\cdots \times F \times E_{j+1} \times \cdots \times E_J) , \quad F \in A_j ,$$

---

[5] The use here of the term 'pseudomeasure' is consistent with its use in harmonic analysis -- see Remark 4.13.

is a signed measure on (X_j, A_j). Observe also this: If we strengthen (4.1) to

$$\sum_{n_1,\ldots,n_J=1}^{\infty} \mu(F_1(n_1) \times \cdots \times F_J(n_J)) = \mu(E_1 \times \cdots \times E_J) ,$$

where the limit $\sum_{n_1,\ldots,n_J=1}^{\infty}$ is taken with respect to finite subsets of $\mathbb{N}^J$ directed by inclusion, then μ has finite 'total variation'

(4.2)

$$\sup\{ \sum_{n_1,\ldots,n_J} |\mu(F_1(n_1) \times \cdots \times F_J(n_J))| : \{F_j(n)\}_{n \in \mathbb{N}} \text{ partition of } X_j, \ j=1,\ldots,J\} =$$

$$= |\mu|(\mathop{\times}_{j=1}^{J} X_j) < \infty .$$

In this case, extending μ in a standard way to a function on $\mathop{\times}_{j=1}^{J} A_j$, we obtain a bona fide signed measure on the product space $\mathop{\times}_{j=1}^{J} X_j$.

'Multilinearizing' the notion of total variation given in (4.2), we define the Fréchet variation of $\mu \in F_J$ over a measurable rectangle $E_1 \times \cdots \times E_J$ as

(4.3) $\quad |\mu|_{F_J}(E_1 \times \cdots \times E_J) = \sup\{\|\sum_{n_1,\ldots,n_J=1}^{N} \mu(F_1(n_1) \times \cdots \times F_J(n_J)) r_{n_1} \otimes \cdots \otimes r_{n_J}\|_\infty :$

$$\{F_j(n)\}_{n \in \mathbb{N}} \text{ partition of } E_j, \ j=1,\ldots,J, \text{ and } N > 0\} .$$

'Countable subadditivity' of the Fréchet variation is an easy consequence, which we formalize as

Lemma 4.2

Let $\mu \in F_J$, $E_j \in A_j$ and $\{F_j(n)\}_{n \in \mathbb{N}}$ be a partition of E_j, $j=1,\ldots,J$. Then

(4.4) $\quad |\mu|_{F_J}(E_1 \times \cdots \times E_J) \leq \sum_{n_1,\ldots,n_J} |\mu|_{F_J}(F_1(n_1) \times \cdots \times F_J(n_J)) .$

Theorem 4.3

$$\|\mu\|_{F_J} \equiv |\mu|_{F_J}(X_1 \times \cdots \times X_J) < \infty \quad \text{for all} \quad \mu \in F_J .$$

The proof of the theorem follows the basic strategy of the argument given in Rudin's book [19] establishing that the total variation of a complex measure is finite (see Lemma 6.3 and Theorem 6.4 in [19]).

Lemma 4.4

Suppose $(a_{j_1 \cdots j_J})^N_{j_1, \ldots, j_J = 1}$ is an arbitrary finitely supported tensor. There exist $S_1, \ldots, S_J \subset \{1, \ldots, N\}$ so that

$$\left| \sum_{(j_1, \ldots, j_J) \in S_1 \times \cdots \times S_J} a_{j_1 \cdots j_J} \right| \geq (\tfrac{1}{4})^J \left\| \sum_{j_1, \ldots, j_J = 1}^N a_{j_1 \cdots j_J} r_{j_1} \otimes \cdots \otimes r_{j_J} \right\|_\infty .$$

Proof (by induction on J)

First, for each j define a function on $[0,1]$

$$\phi_j(\omega) = (r_j(\omega) + 1)/2, \quad \omega \in [0,1 ,$$

which assumes the values 1 and 0 ($\phi_j(\omega)$ is the jth binary digit of $\omega \in [0,1]$).

We start the induction with the case $J = 1$ (this is essentially Lemma 6.3 in [19]): Estimate

(4.5)
$$\sup_\omega \left| \sum_{j=1}^N a_j \phi_j(\omega) \right| \geq \tfrac{1}{2} \left(\left\| \sum_{j=1}^N a_j r_j \right\|_\infty - \left| \sum_{j=1}^N a_j \right| \right) .$$

We can assume

(4.6)
$$\left| \sum_{j=1}^N a_j \right| < \tfrac{1}{2} \left\| \sum_{j=1}^N a_j r_j \right\|_\infty$$

(we are done if (4.6) fails). Therefore, combining (4.6) and (4.5), we deduce the existence of $\omega_0 \in [0,1]$ with the property

$$\left| \sum_{j=1}^N a_j \phi_j(\omega_0) \right| \geq (\tfrac{1}{4}) \left\| \sum_{j=1}^N a_j r_j \right\|_\infty .$$

And so

$$S = \{1 \leq j \leq N : \phi_j(\omega_0) = 1\}$$

is the required set in this case.

We continue the induction: Let $J > 1$ and assume the lemma true in the case $J - 1$. Estimate

$$(4.7) \qquad \sup_{\omega_1,\ldots,\omega_J} | \sum_{j_1,\ldots,j_J=1}^{N} a_{j_1\cdots j_J} \phi_{j_1}(\omega_1)\cdots\phi_{j_J}(\omega_J) | \geq$$

$$\geq (\tfrac{1}{2})^J (\| \sum_{j_1,\ldots,j_J=1}^{N} a_{j_1\cdots j_J} r_{j_1}\otimes\cdots\otimes r_{j_J} \|_\infty$$

$$- \sum_{k=0}^{J-1} \sum_{\substack{T\subset\{1,\ldots,J\} \\ |T|=k \\ T=(i_1,\ldots,i_k) \\ i_1<i_2<\cdots<i_k}} \| \sum_{j_1,\ldots,j_J=1}^{N} a_{j_1\cdots j_J} r_{j_{i_1}}\otimes\cdots\otimes r_{j_{i_k}} \|_\infty) .$$

We can assume that for each $T \subset \{1,\ldots,J\}$, $T = (i_1,\ldots,i_k)$ in the summation above, we have

$$(4.8) \quad \| \sum_{j_1,\ldots,j_J=1}^{N} a_{j_1\cdots j_J} r_{j_1}\otimes\cdots\otimes r_{j_J} \|_\infty \geq 4^{J-k} \| \sum_{j_1,\ldots,j_J=1}^{N} a_{j_1\cdots j_J} r_{j_{i_1}}\otimes\cdots\otimes r_{j_{i_k}} \|_\infty :$$

For, should (4.8) fail for some such T, apply the induction hypothesis and obtain the lemma. Therefore, combining (4.8) and (4.7), we deduce

$$(4.9) \qquad \| \sum_{j_1,\ldots,j_J=1}^{N} a_{j_1\cdots j_J} \phi_{j_1}\otimes\cdots\otimes\phi_{j_J} \|_\infty \geq$$

$$\geq (\tfrac{1}{2})^J \| \sum_{j_1,\ldots,j_J=1}^{N} a_{j_1\cdots j_J} r_{j_1}\otimes\cdots\otimes r_{j_J} \|_\infty (1 - \sum_{k=0}^{J-1} \tfrac{1}{k!} \binom{J-1}{k}/4^{J-k}) .$$

Following elementary computations, we estimate

$$1 - \sum_{k=0}^{J-1} \tfrac{1}{k!} \binom{J-1}{k}/4^{J-k} \geq (\tfrac{1}{2})^J$$

for all $J > 0$, and thus obtain from (4.9) a point $(\omega_1,\ldots,\omega_J)\varepsilon[0,1]^J$ for which

$$| \sum_{j_1,\ldots,j_J=1}^{N} a_{j_1\cdots j_J} \phi_{j_1}(\omega_1)\cdots\phi_{j_J}(\omega_j) | \geq (\tfrac{1}{4})^J \| \sum_{j_1,\ldots,j_J=1}^{N} a_{j_1\cdots j_J} r_{j_1}\otimes\cdots\otimes r_{j_J} \|_\infty .$$

And so, we write

$$S_i = \{1 \le j \le N : \phi_j(\omega_i) = 1\}, \quad i = 1, \ldots, J,$$

which are the required sets in this case. $\boxed{\text{x}}$

Next we formalize a 'Fubini-type' property which follows directly from (4.1):

Lemma 4.5

Let $\mu \in F_J$, $E_j \in A_j$ and $\{F_j(n)\}_{n \in \mathbb{N}}$ be a partition of E_j, $j = 1, \ldots, J$. Then, for every permutation σ of $\{1, \ldots, J\}$

$$(4.10) \quad \sum_{n_{\sigma(1)}} \left(\cdots \left(\sum_{n_{\sigma(J)}} \mu(F_1(n_1) \times \cdots \times F_J(n_J)) \right) \cdots \right) = \mu(E_1 \times \cdots \times E_J).$$

Proof of Theorem 4.3

Suppose the assertion is false, i.e.

$$(4.11) \quad |\mu|_{F_J}(X_1 \times \cdots \times X_J) = \infty.$$

By definition, we then have partitions $\{F_j(n)\}_{n \in \mathbb{N}}$ of X_j, $j = 1, \ldots, J$, so that

$$\left\| \sum_{n_1, \ldots, n_J = 1}^{N} \mu(F_1(n_1) \times \cdots \times F_J(n_J)) r_{n_1} \otimes \cdots \otimes r_{n_J} \right\|_\infty > 4^J \left(1 + |\mu(X_1 \times \cdots \times X_J)| \right).$$

By an application of Lemma 4.4, we obtain $S_1, \ldots, S_J \subset \{1, \ldots, N\}$ so that

$$(4.12) \quad \left| \sum_{(n_1, \ldots, n_J) \in S_1 \times \cdots \times S_J} \mu(F_1(n_1) \times \cdots \times F_J(n_J)) \right| \ge 1 + |\mu(X_1 \times \cdots \times X_J)|.$$

Define

$$H_j = \bigcup_{n \in S_j} F_j(n), \quad j = 1, \ldots, J,$$

and rewrite (4.12) accordingly,

$$(4.13) \quad |\mu(H_1 \times \cdots \times H_J)| \ge 1 + |\mu(X_1 \times \cdots \times X_N)| > 1.$$

Consider now the collection of rectangles

$$R = \{E_1 \times \cdots \times E_J : E_i = H_i \text{ or } \sim H_i, \ i=1,\ldots,J\}.$$

Estimate, by (4.13) and (4.1),

(4.14)
$$\left| \sum_{\substack{R \in R \\ R \neq H_1 \times \cdots \times H_J}} \mu(R) \right| > 1.$$

By Lemma 4.2 and assumption (4.11), we find a rectangle $R_1 \in R$ for which

(4.15)
$$|\mu|_{F_J}(R_1) = \infty.$$

Write

$$R^{(1)} = R \sim \{R_1\}.$$

Starting with (4.15) (in place of (4.11)), we repeat the procedure described above (with R_1 replacing $X_1 \times \cdots \times X_J$), and thus obtain two sequences $(R^{(k)})_{k=1}^{\infty}$ and $(R_k)_{k=1}^{\infty}$. The first is a sequence of finite collections of mutually disjoint rectangles and the second is a sequence of rectangles; together they satisfy the following properties: For each $k \geq 0$,

(4.16)
$$\begin{cases} R_{k+1} \notin R^{(k+1)} \text{ and} \\[2mm] R_k = R_{k+1} \cup \left(\bigcup_{R \in R^{(k+1)}} R \right), \\[2mm] \text{which (by (4.1)) implies} \\[2mm] \mu(R_k) = \mu(R_{k+1}) + \sum_{R \in R^{(k+1)}} \mu(R) \\[2mm] (\text{here } R_0 = X_1 \times \cdots \times X_J); \end{cases}$$

(4.17)
$$\begin{cases} \text{for each } k \geq 1 \text{ either } \left| \sum_{R \in R^{(k)}} \mu(R) \right| > 1 \\[2mm] \text{or } |\mu(R)| > 1 \text{ for some } R \in R^{(k)}. \end{cases}$$

Sublemma

$$\mu(R_k) \to 0 \text{ as } k \to \infty.$$

Proof

Let $Q_1, \ldots, Q_J$ be the canonical projections from $X_1 \times \cdots \times X_J$ onto $X_1, \ldots, X_J$, respectively. For each $j = 1, \ldots, J$ construct a partition of X_j as follows:

$$F_j(1) = X_j \sim Q_j[R_1]$$
$$\vdots \qquad \vdots$$
$$F_j(n) = Q_j[R_{n-1}] \sim Q_j[R_n]$$
$$\vdots \qquad \vdots$$

Observe that for each $k \geq 0$

$$R_k = \bigcup_{n_1, \ldots, n_J = k+1}^{\infty} F_1(n_1) \times \cdots \times F_J(n_J) ,$$

and, by (4.1), deduce

$$(4.18) \qquad \mu(R_k) = \sum_{n_1 = k+1}^{\infty} (\cdots (\sum_{n_J = k+1}^{\infty} \mu(F_1(n_1) \times \cdots \times F_J(n_J))) \cdots) .$$

An application of Lemma 4.5 to (4.18) in the case $k = 0$ implies: For every $\varepsilon > 0$ there exists $N > 0$ so that $|\mu(R_k)| < \varepsilon$ for all $k \geq N$. ☒

A recursive application of (4.16) implies that for each k

$$\mu(X_1 \times \cdots \times X_J) = \sum_{j=1}^{k} (\sum_{R \in \tilde{R}(j)} \mu(R)) + \mu(R_k) .$$

Therefore, by the Sublemma above, we deduce

$$\mu(X_1 \times \cdots \times X_J) = \lim_{k \to \infty} \sum_{j=1}^{k} (\sum_{R \in \tilde{R}(j)} \mu(R))$$

which stands in contradiction with (4.17), and thus establish the theorem. ☒

b. <u>J/K-dimensional Fréchet pseudomeasures</u>

We keep all previous notation: $J \geq K > 0$ are fixed integers, and $(S_\alpha)_{\alpha=1}^{\binom{J}{K}}$ is the collection of all K-subsets of $\{1,\ldots,J\}$ each of which is enumerated $S_\alpha = (\alpha_1,\ldots,\alpha_K)$. Given $\alpha = 1,\ldots,\binom{J}{K}$, define the product space

(4.19)
$$Y_\alpha = \mathop{\times}_{j \varepsilon S_\alpha} X_j ,$$

and the corresponding product σ-algebra

$$\Theta_\alpha = \mathop{\times}_{j \varepsilon S_\alpha} A_j .$$

From now on we view Y_α as the measurable space $(Y_\alpha, \Theta_\alpha)$. As in section 2 (2.9), define the canonical projections

$$P_\alpha : \mathop{\times}_{j=1}^{J} X_j \to Y_\alpha$$

by

$$P_\alpha(x_1,\ldots,x_J) = (x_{\alpha_1},\ldots,x_{\alpha_K}) , \quad \alpha = 1,\ldots,\binom{J}{K} .$$

Given arbitrary measurable sets $E_\alpha \subset Y_\alpha$, $\alpha = 1,\ldots,\binom{J}{K}$, we consider

(4.20)
$$\mathop{\cap}_{\alpha=1}^{\binom{J}{K}} P_\alpha^{-1}[E_\alpha]$$

which we call a measurable generalized rectangle in $\mathop{\times}_{j=1}^{J} X_j$ (to develop some 'visual' intuition examine the case $J = 3$, $K = 2$). Observe that for each $\alpha = 1,\ldots,\binom{J}{K}$, and arbitrary sets $E,E' \subset Y_\alpha$, we have

$$P_\alpha^{-1}[E \cup E'] = P_\alpha^{-1}[E] \cup P_\alpha^{-1}[E']$$

and

$$P_\alpha^{-1}[E \cap E'] = P_\alpha^{-1}[E] \cap P_\alpha^{-1}[E'] .$$

Therefore, we conclude that the collection of generalized rectangles in $\mathop{\times}_{j=1}^{J} X_j$ is closed under the σ-algebra operations in Θ_α for each α.

Definition 4.6

A scalar valued function defined on the measurable generalized
rectangles in $\underset{j=1}{\overset{J}{\times}} X_j$ given by (4.20) is an $F_{J/K}$-pseudomeasure if the
following holds for every $E_\alpha \varepsilon \Theta_\alpha$, $\alpha=1,\ldots,\binom{J}{K}$:

(4.21) For every $\alpha=1,\ldots,\binom{J}{K}$ and every partition

$\{F_\alpha(n)\}_{n\varepsilon\mathbb{N}}$ of $E_\alpha \subset Y_\alpha$,

$$\sum_{n=1}^{\infty} \mu(P_1^{-1}[E_1]\cap\cdots\cap P_\alpha^{-1}[F_\alpha(n)]\cap\cdots\cap P_{\binom{J}{K}}^{-1}[E_{\binom{J}{K}}]) = \mu(\overset{\binom{J}{K}}{\underset{\alpha=1}{\cap}} P_\alpha^{-1}[E_\alpha]) .$$

The space of $F_{J/K}$-pseudomeasures on $\underset{j=1}{\overset{J}{\times}} X_j$ is denoted by
$F_{J/K}(\underset{j=1}{\overset{J}{\times}} X_j) = F_{J/K}$. (We shall refer to F-pseudomeasures when the
'dimension' J/K is not specified.)

Remark 4.7

The requirement (4.21) is equivalent to the following: For all
measurable sets $E_\alpha \subset Y_\alpha$, $\alpha=1,\ldots,\binom{J}{K}$, and each $\beta \varepsilon \{1,\ldots,\binom{J}{K}\}$, the
function defined on Θ_β by

$$\mu(\underset{\substack{1\le\alpha\le\binom{J}{K}\\ \alpha\neq\beta}}{\cap} P_\alpha^{-1}[E_\alpha]\cap P_\beta^{-1}[F]) , \quad F\varepsilon\Theta_\beta ,$$

is a signed measure on Y_β . Indeed, $\mu\varepsilon F_{J/K}(\underset{j=1}{\overset{J}{\times}} X_j)$ can be realized
canonically as an $F_{J/K}$-pseudomeasure $\tilde{\mu}$ defined on $\underset{\alpha=1}{\overset{\binom{J}{K}}{\times}} Y_\alpha$:

(4.22) $\tilde{\mu}(E_1\times\cdots\times E_{\binom{J}{K}}) = \mu(\overset{\binom{J}{K}}{\underset{\alpha=1}{\cap}} P_\alpha^{-1}[E_\alpha]) , \quad E_\alpha \subset Y_\alpha , \quad \alpha=1,\ldots,\binom{J}{K} .$

Guided by the observation above, we proceed to define the
$F_{J/K}$-variation of $\mu\varepsilon F_{J/K}(\underset{j=1}{\overset{J}{\times}} X_j)$ as follows: We start with partitions
$\{F_j(n)\}_{n\varepsilon\mathbb{N}}$ of X_j, $j=1,\ldots,J,$ which give rise to partitions

$$\pi_\alpha = \{F_{\alpha_1}(n_1) \times \cdots \times F_{\alpha_K}(n_K)\}_{n_1,\ldots,n_K \varepsilon \mathbb{N}}$$

of Y_α, $\alpha=1,\ldots,\binom{J}{K}$. Now observe that for every $c_\alpha \varepsilon \pi_\alpha$, $\alpha=1,\ldots,\binom{J}{K}$,

(4.23) $P_1^{-1}[c_1] \cap \cdots \cap P_{\binom{J}{K}}^{-1}[c_{\binom{J}{K}}] = F_1(n_1) \times \cdots \times F_J(n_J)$

$$if \quad P_\alpha[F_1(n_1) \times \cdots \times F_J(n_J)] = c_\alpha$$

$$for \ each \quad \alpha=1,\ldots,\binom{J}{K},$$

$$= \emptyset \quad otherwise.$$

Applying Theorem 4.3 to $\tilde{\mu} \varepsilon F_{\binom{J}{K}}$, we obtain

(4.24) $\left\| \displaystyle\sum_{k_1,\ldots,k_{\binom{J}{K}}} \tilde{\mu}(c_1(k_1) \times \cdots \times c_{\binom{J}{K}}(k_{\binom{J}{K}})) r_{k_1} \otimes \cdots \otimes r_{k_{\binom{J}{K}}} \right\|_\infty < \|\tilde{\mu}\|_{F_{\binom{J}{K}}}$,

where $\{c_\alpha(k)\}_{k \varepsilon \mathbb{N}} = \pi_\alpha$, $\alpha=1,\ldots\binom{J}{K}$. But in view of (4.23),

$$\tilde{\mu}\left(c_1(k_1) \times \cdots \times c_{\binom{J}{K}}(k_{\binom{J}{K}})\right) = \begin{cases} = \mu\left(F_1(n_1) \times \cdots \times F_J(n_J)\right) \\[1em] if \quad P_\alpha[F_1(n_1) \times \cdots \times F_J(n_J)] = c_\alpha \\[0.5em] for \ each \quad \alpha=1,\ldots,\binom{J}{K} \\[1em] = 0 \quad otherwise. \end{cases}$$

Therefore, appropriately indexing the Rademacher functions by $\mathbb{N}^K$ (as in (3.1)), we obtain from Theorem 4.3 and (4.24)

Theorem 4.8

For all $\mu \varepsilon F_{J/K}$,

$$\|\mu\|_{F_{J/K}} \equiv \sup\{ \left\| \sum_{\substack{n_1,\ldots,n_J=1 \\ n=(n_1,\ldots,n_J)}}^{N} \mu(F_1(n_1) \times \cdots \times F_J(n_J)) r_{P_1(n)} \otimes \cdots \otimes r_{P_{\binom{J}{K}}(n)} \right\|_\infty :$$

$$\{F_j(n)\}_{n \varepsilon \mathbb{N}} \ partition \ of \ X_j, \ j=1,\ldots,J \ ; \ N > 0\}$$

$$\leq \ \|\tilde{\mu}\|_{F_{\left(\begin{smallmatrix}J\\K\end{smallmatrix}\right)}} \ < \ \infty \ .$$

c. Integration with respect to F-pseudomeasures

A theory of integration in the present framework follows naturally the theory of integration with respect to ordinary signed measures (elements of F_1). We denote the Banach algebra of bounded measurable functions on a measurable space X by $L^\infty(X)$, which we equip with the usual supremum-norm. Given $f_1 \in L^\infty(X_1), \ldots, f_J \in L^\infty(X_J)$ and $\mu \in F_J$, our aim is to define by induction the integral

$$\int_{\underset{j=1}{\overset{J}{\times}} X_j} f_1 \otimes \cdots \otimes f_J \, d\mu \ .$$

Assume now that $J \geq 2$ and let μ_{f_1} be the function on $A_2 \times \cdots \times A_J$ given by

$$(4.25) \quad \mu_{f_1}(E_2 \times \cdots \times E_J) = \int_{X_1} f_1(x)\mu(dx \times E_2 \times \cdots \times E_J), \quad E_2 \in A_2, \ldots, E_J \in A_J \ .$$

(The right hand side of (4.25) is an ordinary Lebesgue integral with respect to the signed measure $\mu(\cdot \times E_2 \times \cdots \times E_J)$.)

Lemma 4.9

$$(4.26) \qquad\qquad \mu_{f_1} \in F_{J-1}\left(\underset{j=2}{\overset{J}{\times}} X_j\right)$$

and

$$(4.27) \qquad\qquad \|\mu_{f_1}\|_{F_{J-1}} \leq \|f_1\|_\infty \|\mu\|_{F_J} \ .$$

Proof

By standard convergence theorems, it suffices to check the lemma for simple functions

$$f_1 = f = \sum_i a_i \chi_{F_i}, \quad F_i \cap F_j = \emptyset \ \text{if} \ i \neq j \ .$$

(4.26) and (4.27) are proved by induction on J: Let $J = 2$ and suppose
that $\{E_j\}_{j \in \mathbb{N}}$ is a partition of X_2. We have

(4.28)
$$\mu_f \left(\bigcup_j E_j \right) = \sum_i a_i \mu \left(\left(\bigcup_j E_j \right) \times F_i \right)$$

$$= \sum_i a_i \sum_j \mu (E_j \times F_i) \qquad \text{(by (4.1)}.$$

We require the following

Sublemma

Suppose $\sup_N \left\| \sum_{i,j=1}^{N} b_{ij} r_i \otimes r_j \right\|_\infty \leq 1.$ Then

(4.29)
$$\lim_{N \to \infty} \left\| \sum_{i,j=N}^{\infty} b_{ij} r_i \otimes r_j \right\|_\infty = 0 .$$

Proof

To establish (4.29), assume that it fails and apply Lemma 4.4
(repeatedly) to obtain a contradiction with the supposition.   ☒

By Theorem 4.3,

$$\left\| \sum_{i,j} \mu (E_j \times F_i) r_i \otimes r_j \right\|_\infty \leq \| \mu \|_{F_2} .$$

Therefore, by (4.29), we can interchange the summations in (4.28)

$$\sum_i a_i \sum_j \mu (E_j \times F_i) = \sum_j \sum_i a_i \mu (E_j \times F_i)$$

$$= \sum_j \mu_f (E_j) ,$$

and obtain

$$\left\| \sum_j \mu_f (E_j) r_j \right\|_\infty \leq \sup_j |a_j| \, \| \mu \|_{F_2} ,$$

thus verifying the lemma in the case $J = 2$. The proof of the general
inductive step is based on the case $J = 2$ and a multidimensional
version of the sublemma above -- details are omitted.   ☒

Under Lemma 4.9 we define integration with respect to $\mu \in F_J$ by induction

on J (the case $J = 1$ is, of course, ordinary Lebesgue integration):

$$\int\limits_{\substack{\times X_j \\ j=1}}^{J} f_1 \otimes \cdots \otimes f_J \, d\mu = \int\limits_{\substack{\times X_j \\ j=2}}^{J} f_2 \otimes \cdots \otimes f_J \, d\mu_{f_1} \, .$$

We formalize (omitting the proof by induction)

Proposition 4.10

(4.30)
$$\left| \int\limits_{\substack{\times X_j \\ j=1}}^{J} f_1 \otimes \cdots \otimes f_J \, d\mu \right| \leq \| f_1 \|_\infty \cdots \| f_J \|_\infty \| \mu \|_{F_J}$$

and

$$\int\limits_{\substack{\times X_j \\ j=1}}^{J} f_1 \otimes \cdots \otimes f_J \, d\mu = \int\limits_{\substack{\times X_j \\ j=1}}^{J} f_{\sigma(1)} \otimes \cdots \otimes f_{\sigma(J)} \, d\mu$$

for all $\mu \in F_J$, $f_1 \in L^\infty(X_1), \ldots, f_J \in L^\infty(X_J)$, and permutations σ of

$\{1, \ldots, J\}$.

Integration with respect to the 'fractional' F-pseudomeasures given

in Definition 4.6 is carried out via the realization of $\mu \in F_{J/K}\left(\overset{J}{\underset{j=1}{\times}} X_j \right)$

as an $F_{\binom{J}{K}}$-pseudomeasures $\tilde{\mu}$ in $F_{\binom{J}{K}}\left(\overset{\binom{J}{K}}{\underset{\alpha=1}{\times}} Y_\alpha \right)$ (see Remark 4.7). The

archtypical function f on $\overset{J}{\underset{j=1}{\times}} X_j$ that is integrable with respect to

$\mu \in F_{J/K}$ is this: let $g_1, \ldots, g_{\binom{J}{K}}$ be bounded measurable functions on

$Y_1, \ldots, Y_{\binom{J}{K}}$, respectively, and define

(4.31) $f(x_1, \ldots, x_J) = g_1\left(P_1(x_1, \ldots, x_J)\right) \cdots g_{\binom{J}{K}}\left(P_{\binom{J}{K}}(x_1, \ldots, x_J)\right) .$

The integral of f with respect to μ is given by

(4.32)
$$\int\limits_{\substack{J \\ \times X_j \\ j=1}} f d\mu = \int\limits_{\substack{\binom{J}{K} \\ \times Y_\alpha \\ \alpha=1}} g_1 \otimes \cdots \otimes g_{\binom{J}{K}} \, d\tilde{\mu} \, .$$

We proceed to establish that the left hand side of (4.32) is well defined.

<u>Proposition 4.11</u>

Suppose $g_1, \ldots, g_{\binom{J}{K}}$ are bounded measurable functions on $Y_1, \ldots, Y_{\binom{J}{K}}$, respectively, so that

$$g_1\big(P_1(x)\big) \cdots g_{\binom{J}{K}}\big(P_{\binom{J}{K}}(x)\big) = 0 \quad \text{for all} \quad x \in \underset{j=1}{\overset{J}{\times}} X_j \, .$$

Then, for all $\mu \in F_{J/K}\big(\underset{j=1}{\overset{J}{\times}} X_j\big)$

$$\int\limits_{\substack{\binom{J}{K} \\ \times Y_\alpha \\ \alpha=1}} g_1 \otimes \cdots \otimes g_{\binom{J}{K}} \, d\tilde{\mu} = 0 \, .$$

<u>Proof</u>

To illustrate ideas, we shall argue the case $J = 3$, $K = 2$ (the argument in the general case is similar):

Suppose f, g, h are bounded measurable functions on

$$Y_1 = X_1 \times X_2, \ Y_2 = X_2 \times X_3, \ Y_3 = X_1 \times X_3 \, .$$

respectively, so that $\|f\|_\infty = \|g\|_\infty = \|h\|_\infty = 1$ and

(4.33) $f(x_1, x_2) g(x_2, x_3) h(x_1, x_3) = 0$ for all $(x_1, x_2, x_3) \in X_1 \times X_2 \times X_3$.

Given an arbitrary $\varepsilon > 0$, approximate f, g, h by simple functions

$$\phi_1 = \sum_i a_i \chi_{A_i} \, ,$$

$$\phi_2 = \sum_i b_i \chi_{B_i} \, ,$$

$$\phi_3 = \sum_i c_i \chi_{C_i} \, ,$$

where $\{A_i\}_i$, $\{B_i\}_i$, $\{C_i\}_i$ are partitions of Y_1, Y_2, Y_3, respectively,

and

(4.34) $$\| f - \phi_1 \|_\infty , \quad \| g - \phi_2 \|_\infty , \quad \| h - \phi_3 \|_\infty < \varepsilon .$$

By (4.30) in Proposition 4.10, it suffices to prove

(4.35) $$\left| \int_{Y_1 \times Y_2 \times Y_3} \phi_1 \otimes \phi_2 \otimes \phi_3 \, d\tilde{\mu} \right| < \varepsilon \| \tilde{\mu} \|_{F_{\binom{3}{2}}} .$$

Write (by (4.22))

(4.36) $$\int_{Y_1 \times Y_2 \times Y_3} \phi_1 \otimes \phi_2 \otimes \phi_3 \, d\tilde{\mu} = \sum_{i,j,k} a_i b_j c_k \, \mu(P_1^{-1}[A_i] \cap P_2^{-1}[B_j] \cap P_3^{-1}[C_k])$$

and observe the following:

Either there is a point

(4.37) $$(x_1^{(i)}, x_2^{(j)}, x_3^{(k)}) \in P_1^{-1}[A_i] \cap P_2^{-1}[B_j] \cap P_3^{-1}[C_k]$$

in which case the corresponding summand in (4.36) can be written with

(4.38)
$$a_i b_j c_k = (f(x_1^{(i)}, x_2^{(j)}) + \varepsilon_{ij}^{(1)})(g(x_2^{(j)}, x_3^{(k)}) + \varepsilon_{jk}^{(2)})(h(x_1^{(i)}, x_3^{(k)}) + \varepsilon_{ik}^{(3)})$$

where $| \varepsilon_{ij}^{(1)} |, | \varepsilon_{jk}^{(2)} |, | \varepsilon_{ik}^{(3)} | < \varepsilon$ (by (4.34)),

or

$$P_1^{-1}[A_i] \cap P_2^{-1}[B_j] \cap P_3^{-1}[C_k] = \emptyset$$

in which case the corresponding summand in (4.36) vanishes. Write

$$F = \{(i,j,k) : P_1^{-1}[A_i] \cap P_2^{-1}[B_j] \cap P_3^{-1}[C_k] \neq \emptyset\}$$

and, following (4.37) note that

$(i,j,k) \in F$ if and only if $(x_1^{(i)}, x_2^{(j)}) \in A_i$, $(x_2^{(j)}, x_3^{(k)}) \in B_j$, $(x_1^{(i)}, x_3^{(k)}) \in C_k$.

Therefore, we obtain from Theorem 4.3

(4.39) $$\left\| \sum_{(i,j,k) \in F} \rho_{ijk} \, r_{ij} \otimes r_{jk} \otimes r_{ik} \right\|_\infty \leq \| \tilde{\mu} \|_{F_{\binom{3}{2}}} ,$$

where

$$\rho_{ijk} = \mu(P_1^{-1}[A_i] \cap P_2^{-1}[B_j] \cap P_3^{-1}[C_k])$$

and $R = \{r_{k_1 k_2}\}_{k_1, k_2 \in \mathbb{N}}$ is the Rademacher system indexed by $\mathbb{N}^2$, as in (3.1) with $K = 2$ (the argument here is analogous to the one leading from Theorem 4.3 via (4.23) and (4.24) to Theorem 4.8).

In view of (4.38), rewrite (4.36) as

$$\sum_{(i,j,k) \in F} \rho_{ijk} (f(x_1^{(i)}, x_2^{(j)}) + \varepsilon_{ij}^{(1)})(g(x_2^{(j)}, x_3^{(k)}) + \varepsilon_{jk}^{(2)})(h(x_1^{(i)}, x_3^{(i)}) + \varepsilon_{ik}^{(3)}).$$

Finally, multiplying out the three factors in the summand above and applying (4.33) and (4.39), we obtain

$$\left| \int_{Y_1 \times Y_2 \times Y_3} \phi_1 \otimes \phi_2 \otimes \phi_3 \, d\tilde{\mu} \right| < 3\varepsilon(1+\varepsilon)^2 \|\tilde{\mu}\|_F \quad \binom{3}{2}$$

which (modulo an insignificant factor) establishes the required (4.35).

$\boxed{\text{x}}$

d. <u>Bounded multilinear forms on $C_o(x)$</u>

Fréchet's theorem [6] about bounded bilinear forms on $C([0,1])$ -- indeed, the historical catalyst for this paper -- can be neatly reformulated and extended in the present framework. Let $X_1, \ldots, X_J$ be locally compact Hausdorff spaces and $B_1, \ldots, B_J$ be the Borel fields generated by the topologies in $X_1, \ldots, X_J$, respectively. $C_o(X)$ will denote the Banach algebra of continuous functions vanishing at infinity on a locally compact X. We consider now the J-fold projective tensor product

(4.40)
$$V_J(\underset{j=1}{\overset{J}{\times}} X_j) = V_J = C_o(X_1) \hat{\otimes} \cdots \hat{\otimes} C_o(X_J)$$

$$= \{f \in C_o(\underset{j=1}{\overset{J}{\times}} X_j) : (*) \ f = \sum_k a_k g_k^{(1)} \otimes \cdots \otimes g_k^{(J)} \quad \text{where}$$

$$\sum_k |a_k| < \infty \quad \text{and} \quad \|g_k^{(j)}\|_\infty \le 1 \quad \text{for each} \quad j=1,\ldots,J \quad \text{and all} \quad k\}.$$

Normed by

$$\|f\|_{V_J} = \inf\{\sum_k |a_k| \; : \; f \text{ given by } (*) \text{ in } (4.40)\} \, ,$$

V_J is a Banach algebra with pointwise multiplication. A scalar valued function γ defined on $C_o(X_1) \times \cdots \times C_o(X_J)$ is said to be a bounded J-linear form if it is linear in each of its J coordinates and (its norm)

$$\|\gamma\| = \sup\{|\gamma(f_1,\ldots,f_J)| \; : \; f_j \text{ in the unit ball of } C_o(X_j), \; j=1,\ldots,J\}$$

is finite. The dual space of the projective tensor algebra V_J is precisely the space of bounded J-linear forms, described by the following multilinear Riesz Representation Theorem:

Theorem 4.12

V_J^* is canonically isomorphic to the space of regular F_J-pseudomeasures on $\underset{j=1}{\overset{J}{\times}} X_j$ (normed by the Fréchet variation $\| \; \|_{F_J}$): The identification

$$\gamma \longleftrightarrow \mu^{(\gamma)} \, , \quad \gamma \in V_J^* \, , \quad \mu^{(\gamma)} \in F_J \, ,$$

is given by

(4.41)
$$\gamma(f_1,\ldots,f_J) = \int_{\underset{j=1}{\overset{J}{\times}} X_j} f_1 \otimes \cdots \otimes f_J \, d\mu^{(\gamma)} \, ,$$

and satisfies

$$\|\gamma\| = \|\mu^{(\gamma)}\|_{F_J} \, .$$

(Regularity of F_J-pseudomeasures is defined in the obvious way.)

Proof (by induction)

The case $J = 1$ is the Riesz Representation Theorem (e.g., Theorem 6.19 in [19]). Assume the case $J-1$, $J>1$. Let γ be a bounded linear functional on V_J, and fix an arbitrary $f \in C_o(X_J)$. The action

$$f_1 \otimes \cdots \otimes f_{J-1} \rightarrow \gamma(f_1,\ldots,f_{J-1},f), \quad f_j \in C_o(X_j), \; j=1,\ldots,J-1 \, ,$$

defines a bounded linear functional on $\quad V_{J-1}(\overset{J-1}{\underset{j=1}{\times}} X_j)$ and therefore, by

the induction hypothesis, we obtain $\quad \mu_f \in F_{J-1}(\overset{J-1}{\underset{j=1}{\times}} X_j)$ so that

(4.42) $\qquad \gamma(f_1,\ldots,f_{J-1},f) = \int\limits_{\substack{J-1 \\ \underset{j=1}{\times} X_j}} f_1 \otimes \cdots \otimes f_{J-1} \, d\mu_f \, .$

Fixing $E_1 \in B_1,\ldots, E_{J-1} \in B_{J-1}$, observe that

$$f \to \mu_f(E_1 \times \cdots \times E_{J-1}), \quad f \in C_o(X_J) \, ,$$

defines a bounded linear functional on $C_o(X_J)$. Therefore, we obtain by

the Riesz Representation Theorem a Borel measure $\mu_{E_1 \times \cdots \times E_{J-1}}$ on X_J

so that

(4.43) $\qquad \mu_f(E_1 \times \cdots \times E_{J-1}) = \int\limits_{X_J} f \, d\mu_{E_1 \times \cdots \times E_{J-1}}$

Finally, we define $\mu^{(\gamma)} \in F_J$ by

$$\mu^{(\gamma)}(E_1 \times \cdots \times E_J) = \mu_{E_1 \times \cdots \times E_{J-1}}(E_J) \, ,$$

and obtain (4.41) from (4.42) and (4.43).

To deduce $\|\gamma\| = \|\mu^{(\gamma)}\|_{F_J}$, we apply (4.30) and follow the

induction outlined above. $\qquad \boxed{\text{x}}$

Remark 4.13

Fréchet's theorem regarding bounded bilinear forms on $C([0,1])$ is,

of course, Theorem 4.12 with $X_1 = X_2 = [0,1]$ given the usual Borel field.

In this case, Fréchet worked with the 'distribution function' of

$\mu \in F_2([0,1]^2)$

$$\phi_\mu(x,y) = \mu([0,x) \times [0,y)), \quad x,y \in [0,1]$$

(see (1.2) in section 1). In subsequent studies (e.g., [14], [15])

M. Morse and W. Transue investigated further the (two-dimensional) Fréchet

variation in the classical setting $[0,1] \times [0,1]$, and a general
topological framework $X_1 \times X_2$, in which case Morse dubbed bounded
bilinear forms on $C_0(X_1) \times C_0(X_2)$ bimeasures.

The projective tensor product V_J was studied extensively in a
context of harmonic analysis by N. Varopoulos [11] (and hence the V). In
particular, when $X_1,\ldots,X_J$ are compact, Varopoulos demonstrates in [11]
that V_J is canonically isomorphic to the restriction algebra
$A(K_1 \times \cdots \times K_J)$ where each K_j, $j=1,\ldots,J$, is a Kronecker set in a compact
abelian group G. Therefore, V_J^* is canonically isomorphic to the space
of 'pseudomeasures' on G^J, elements of $\ell^\infty(\Gamma^J)$, supported in $K_1 \times \cdots \times K_J$
(here 'pseudomeasures' and their support are taken in the usual sense of
harmonic analysis). Theorem 4.12 implies that these 'pseudomeasures,'
elements of $A(K_1 \times \cdots \times K_J)^*$, are the F_J-pseudomeasures in the present
context.

We proceed now to the fractional projective tensor product which we
define as

(4.44)
$$V_{J/K}\left(\mathop{\times}_{j=1}^{J} X_j\right) = V_{J/K}$$

$$= \left\{ f \in C_0\left(\mathop{\times}_{j=1}^{J} X_j\right) : (*) \; f(x) = \sum_i a_i g_i^{(1)}(P_1(x)) \cdots g_i^{\left(\binom{J}{K}\right)}\left(P_{\binom{J}{K}}(x)\right), \right.$$

where $x \in \mathop{\times}_{j=1}^{J} X_j$, $\sum_i |a_i| < \infty$, and $g_i^{(\alpha)}$ is in the unit ball

of $C_0(Y_\alpha)$ for each $\alpha = 1,\ldots,\binom{J}{K}$ and all $i \}$.

Write

$$H_{J/K} = \left\{ (P_1(x),\ldots,P_{\binom{J}{K}}(x)) : x \in \mathop{\times}_{j=1}^{J} X_j \right\},$$

and observe that $V_{J/K}$ can be realized as a quotient of

$C_0(Y_1) \hat{\otimes} \cdots \hat{\otimes} C_0(Y_{\binom{J}{K}})$ by the space

$$Z = \{f \in C_o(Y_1) \hat{\otimes} \cdots \hat{\otimes} C_o(Y_{\binom{J}{K}}) : f \equiv 0 \quad \text{on} \quad H_{J/K}\} .$$

The norm in $V_{J/K}$ is the resulting quotient norm which can be obtained directly also as

$$\|f\|_{V_{J/K}} = \inf\{\sum_i |a_i| : f \quad \text{given by} \quad (*) \quad \text{in} \quad (4.44)\} .$$

Observe now that $F_{J/K}$, via its realization in $F_{\binom{J}{K}}$ (as per (4.22) and (4.32), is the annhilator of Z (e.g., Proposition 4.11), and thus obtain

Theorem 4.14

The dual space of $V_{J/K}$ is canonically isometric to the Banach space consisting of all regular $F_{J/K}$-pseudomeasures (normed by $\| \ \|_{F_{J/K}}$) whose action on $V_{J/K}$ is given by integration (4.32).

5. DIMENSION OF SETS AND THE VARIATIONS OF $F_{J/K}$-PSEUDOMEASURES

We shall now examine the Fréchet pseudomeasures of the previous section in the light of a measurement of combinatorial dimension of sets in a Borel measurable framework. In the interest of concreteness, all the work here will be done in the J-fold Cartesian product of $[0,1]$. First, we recast in this setting some notions from section 2. A regular partition of $[0,1]$ will be a countable collection of mutually disjoint intervals whose union is $[0,1]$. The size of a regular partition τ will be measured by

$$\|\tau\| = \sup\{\text{length}(I) \; : \; I\epsilon\tau\}.$$

A grid τ of $[0,1]^J$ will mean here a J-fold Cartesian product of regular partitions of $[0,1]^J$,

$$\tau = \tau_1 \times \cdots \times \tau_J$$

whose size is measured by

$$\|\tau\| = \max\{\|\tau_j\| \; : \; j=1,\ldots J\}.$$

Elements of a grid will be called cells and viewed as subsets of $[0,1]^J$,

$$\tau \ni (F_1,\ldots,F_J) \longleftrightarrow F_1 \times \cdots \times F_J \subset [0,1]^J.$$

Given an arbitrary subset $F \subset [0,1]^J$ and a grid τ, denote

(5.1) $$F_\tau = \{c\epsilon\tau \; : \; c \cap F \neq \emptyset\}.$$

As in section 2, define

(5.2) $$d_{F_\tau}(a) = \sup_s \frac{\Psi_{F_\tau}(s)}{s^a}, \quad 0 < a < \infty,$$

and

(5.3)
$$D_F(a) = \sup_{\substack{\varepsilon>0 \\ \|\tau\|<\varepsilon}} (\inf_{\text{grid } \tau} d_{F_\tau}(a))$$

$$= \lim_{\varepsilon\to 0} (\inf_{\|\tau\|<\varepsilon} d_{F_\tau}(a)).$$

Definition 5.1

The dimension of $F \subset [0,1]^J$ (relative to the Borel structure) is

$$\text{Dim}F = \inf\{a : D_F(a) < \infty\}.$$

$\text{Dim}F$ is said to be exact if $D_F(\text{Dim}F) < \infty$ and asymptotic if $D_F(\text{Dim}F) = \infty$.

The two basic propositions below follow from definitions:

Proposition 5.2

(i) Suppose $F, F' \subset [0,1]^J$. Then

$$\text{Dim}(F \cup F') = \max(\text{Dim}F, \text{Dim}F').$$

(ii) Suppose $F \subset [0,1]^J$, $F' \subset [0,1]^{J'}$. Then

$$\text{Dim}(F \times F') \leq \text{Dim}F + \text{Dim}F'.$$

Proof

(i) Observe that for any grid τ,

$$\Psi_{(F \cup F')_\tau}(s) \leq 2 \max(\Psi_{F_\tau}(s), \Psi_{F'_\tau}(s)) \qquad \text{for all } s > 0.$$

Therefore,

$$d_{(F \cup F')_\tau}(s) \leq 2 \max(d_{F_\tau}(a), d_{F'_\tau}(a)),$$

and hence

$$D_{F \cup F'}(a) \leq 2 \max(D_F(a), D_{F'}(a)),$$

which implies

$$\mathrm{Dim}(F \cup F') \leq \max(\mathrm{Dim}F, \mathrm{Dim}F') .$$

To obtain the reverse inequality, note that

$$\mathrm{Dim}F \geq \mathrm{Dim}E \quad \text{whenever} \quad F \supset E .$$

(ii) Observe that for τ and τ', grids of $[0,1]^J$ and $[0,1]^{J'}$, respectively, we have

$$\Psi_{(F \times F')_{\tau \times \tau'}}(s) = \Psi_{F_\tau}(s) \cdot \Psi_{F'_{\tau'}}(s) \quad \text{for all} \quad s > 0 .$$

Therefore, for every $a, a' > 0$,

$$d_{(F \times F')_{\tau \times \tau'}}(a+a') \leq d_{F_\tau}(a) \cdot d_{F'_{\tau'}}(a')$$

implying the assertion.

$$\boxed{\mathrm{x}}$$

Remark 5.3

In Proposition 5.2(ii), certain hypotheses on the growth of $\Psi_{F_\tau}(s)$ and $\Psi_{F'_{\tau'}}(s)$ as $s \to \infty$ will guarantee that $\mathrm{Dim}(F \times F') = \mathrm{Dim}F + \mathrm{Dim}F'$; these matters will not be pursued here.

Proposition 5.4

Suppose $F \subset [0,1]^J$ is a closed countable set. Then

$$\mathrm{Dim}F = \mathrm{dim}F ,$$

where $\mathrm{dim}F$ is the combinatorial dimension given in Definition 2.1.

Proof

First, we show that if $d_F(a) = \infty$ then $D_F(a) = \infty$ (d_F is given by (2.2)). To this end, let $M > 0$ be arbitrary and note that $d_F(a) = \infty$ means: There are $s > 0$ and $A_1, \ldots, A_J \subset [0,1]$ whose cardinality equals s and

$$|A_1 \times \cdots \times A_J \cap F| > Ms^a .$$

But then, there is a (sufficiently small) $\varepsilon > 0$ so that for every grid τ

of $[0,1]^J$, $\|\tau\| < \varepsilon$, we have $\Psi_{F_\tau}(a) > Ms^a$, which implies $D_F(a) > M$.

Conversely, suppose $d_F(a) < \infty$. Let $Q_1, \ldots, Q_J$ be the canonical projections from $[0,1]^J$ onto $[0,1]$. For each $j=1, \ldots, J$, $Q_j[F]$ is a closed countable set whose complement in $[0,1]$ is a union of countably many mutually disjoint open segments; let τ_j be the regular partition of $[0,1]$ which consists of these open segments and $Q_j[F]$. Write $\tau = \tau_1 \times \cdots \times \tau_J$, and deduce that $d_{F_\tau}(a) = d_F(a)$. But, passing to further refinements of τ_j for each $j=1, \ldots, J$, we can assume that $\|\tau\|$ is as small as we wish, and so we conclude that $D_F(a) \leq d_F(a) < \infty$.

We therefore have: $d_F(a) < \infty$ if and only if $D_F(a) < \infty$, and thus $\dim F = \mathrm{Dim}\, F$.

$\boxed{\text{x}}$

Examples 5.5

(i) The requirement that the countable set $F \subset [0,1]^J$ be closed is necessary in the statement of Proposition 5.4: Let $(\tau^{(n)})_{n=1}^{\infty}$ be a sequence of finite grids in $[0,1]^2$ (grids that contain finitely many cells). Assume that in every $\tau^{(n)}$ each cell has non-empty interior, and that

$$(5.4) \qquad \lim_{n \to \infty} \|\tau^{(n)}\| = 0 .$$

As usual, Q_1 and Q_2 denote the two canonical projections from $[0,1]^2$ onto $[0,1]$. Start with a finite set $F^{(1)}$ in $[0,1]^2$ so that

$$(5.5) \qquad Q_i : F^{(1)} \to [0,1] \quad \text{is one-one,} \quad i=1,2 ,$$

and

$$(5.6) \qquad \text{each cell of } \tau^{(1)} \text{ meets } F^{(1)} .$$

Continue by induction: Let $n > 1$ and select a finite set $F^{(n)}$ in $[0,1]^2$ so that

(5.7) Q_i : $F^{(1)} \cup \cdots \cup F^{(n)} \rightarrow [0,1]$ is one-one,   i=1,2,

and

(5.8)   each cell of $\tau^{(n)}$ meets $F^{(n)}$.

Write

$$F = \bigcup_{n=1}^{\infty} F^{(n)} .$$

Observe that (5.4) and (5.8) imply that F is dense in $[0,1]^2$, and
therefore DimF = 2. On the other hand, the requirement (5.7) guarantees

$$\Psi_F(s) = s \quad \text{for all} \quad s > 0 ,$$

and therefore dimF = 1.

(ii) Following the random constructions of [5], one can obtain in
$[0,1]^J$ closed countable sets of any prescribed dimension. In fact, for
every $1 < \alpha < J$, these random constructions can be 'blown up' and
iterated to produce Cantor-like perfect sets $F \subset [0,1]^J$ with the property
that

$$\text{Dim}(F \cap V) = \alpha$$

for every open V which meets F. The details of such constructions will
appear elsewhere.

(iii) Let $f:[0,1]^2 \rightarrow [0,1]$ be a continuous function, and consider
the 'surface'[6]

$$S = \{(x,y,f(x,y)) : (x,y) \in [0,1]^2\} \subset [0,1]^3 .$$

Given an arbitrary $\varepsilon > 0$, let π be any finite regular partition of
$[0,1]$, $\|\pi\| < \varepsilon$. By the continuity of f, we can find a grid τ of
$[0,1]^2$, $\|\tau\| < \varepsilon$, so that each cell of τ is mapped by f into at most
one element of π. We thus have

$$d_{S_{\tau \times \pi}}(2) = 1 \quad \text{with} \quad \|\tau \times \pi\| < \varepsilon$$

_____

[6] Compare with p. 137 in [1].

and since $\varepsilon > 0$ is arbitrary,

$$D_S(2) \leq 1 .$$

In the other direction, it is easy to verify that $D_S(a) = \infty$ for every $a < 2$. We therefore conclude

$$\text{Dim} S = 2 \quad \text{exactly.}$$

In general, fix $J > L \geq 1$ and let

$$f_1, \ldots, f_{J-1} : [0,1]^L \rightarrow [0,1]$$

be continuous functions. Consider the 'surface'

$$S^{(L)} \doteq \{(x_1, \ldots, x_L, f_1(x_1, \ldots, x_L), \ldots, f_{J-L}(x_1, \ldots, x_L)) : (x_1, \ldots, x_L) \in [0,1]^L\} ,$$

and note that

$$\text{Dim} S^{(L)} = L \quad \text{exactly.}$$

Observe that

$$\dim S^{(L)} = L \quad \text{exactly.}$$

I do not have examples of closed sets $F \subset [0,1]^J$ for which $\text{Dim} F > \dim F$.

We proceed to the connection between F-pseudomeasures on $[0,1]^J$ and the dimension of subsets in $[0,1]^J$, analogous to the connection between bounded multilinear forms on c_o and combinatorial dimension of subsets in $\mathbb{N}^J$ (Theorem 3.1). Suppose $\mu \in F_{J/K}([0,1]^J)$, and fix $p \in (0,\infty)$. Define the p-variation of μ over a subset $F \subset [0,1]^J$ as

$$(5.9) \qquad |\mu|^p(F) = \sup_{\varepsilon > 0} \inf_{\|\tau\| < \varepsilon} \sum_{c \in F_\tau} |\mu(c)|^p ,$$

where τ above denotes a grid of $[0,1]^J$. To fix ideas in the present context, observe the following: First, recall that at the outset $\mu \in F_{J/K}$ is defined only on the generalized rectangles in $[0,1]^J$ (given by (4.20)). If $p = 1$ and $F = [0,1]^J$ then (5.9) becomes

$$(5.10) \qquad |\mu|([0,1]^J) = \sup_{\text{grid } \tau} \sum_{c \in \tau} |\mu(c)| ,$$

which is the usual total variation of μ. And so, if $|\mu|([0,1]^J) < \infty$
then μ can be extended by standard methods to a Borel measure on $[0,1]^J$
whose total variation is given by (5.10). In general, the p-variation
defined by (5.9) gives rise to a 'Hausdorff-type' measure and a
corresponding 'Hausdorff-type' dimension of $F \subset [0,1]^J$ defined as

(5.11) $H\text{-Dim}_\mu F = \inf\{p: |\mu|^p(F) < \infty\}$. [7]

Our aim, guided by the 'discrete' results of section 3, is to show that
the p-variation of an F-pseudomeasure over $F \subset [0,1]^J$ is controlled
precisely by the dimension of F:

Theorem 5.6

Let $\mu \in F_{J/K}([0,1]^J)$, and suppose $F \subset [0,1]^J$. Write

(5.12) $p = \max\{1, 2/(1 + K/\text{Dim}F)\}$.

If DimF is exact then

(5.13) $|\mu|^p(F) \leq \beta\|\mu\|_{F_{J/K}}$.

If DimF is asymptotic then

(5.14) $|\mu|^r(F) \leq \beta_r\|\mu\|_{F_{J/K}}$ for all $r > p$

(β, $\beta_r > 0$ are constants independent of μ).
In particular, for all $\mu \in F_{J/K}([0,1]^J)$

                         $H\text{-Dim}_\mu F \leq p$.

Proof

Without loss of generality assume $\|\mu\|_{F_{J/K}} = 1$. To establish (5.13)
and (5.14), it suffices to prove the following:

---

[7] C. Tricot indicated to me that a similar notion of 'p-variation' of a
    positive measure on $[0,1]$ and a subsequent notion of a 'Hausdorff
    type' dimension are discussed in [1], pp. 139-141.

Claim

Suppose

(5.15) $D_F(a) < \zeta$, $0 < a < \infty$.

Then:

(5.16) $\begin{cases} |\mu|^{2/(1+K/a)}(F) \leq \zeta_a & \text{if } a \geq K \\[2mm] |\mu|(F) \leq \zeta_K & \text{if } a \leq K \end{cases}$

where the (respectively subscripted) ζ's above are positive constants independent of μ.

Proof of Claim: From the definition of $D_F(a)$ and (5.15), we have for every $\varepsilon > 0$ a grid τ of $[0,1]^J$ so that

(5.17) $d_{F_\tau}(a) < \zeta$, $\|\tau\| < \varepsilon$

By Theorem 4.8, we have

$$\left\| \sum_{\substack{F_1(n_1) \times \cdots \times F_J(n_J) \varepsilon \tau \\ n=(n_1,\ldots,n_J)}} \mu(F_1(n_1) \times \cdots \times F_J(n_J)) r_{P_1}(n) \otimes \cdots \otimes r_{P_{\binom{J}{K}}}(n) \right\|_\infty < \|\mu\|_{F_{J/K}},$$

where $\tau = \tau_1 \times \cdots \times \tau_J$, $\tau_1 = \{F_1(n)\}_{n \in \mathbb{N}}, \ldots, \tau_J = \{F_J(n)\}_{n \in \mathbb{N}}$, and the Rademacher system is appropriately indexed by $\mathbb{N}^K$. Therefore, by definition (5.2) of d_{F_τ}, (5.17), and Lemma 3.5, we deduce

(5.18) $\begin{cases} \sum_{c \varepsilon F_\tau} |\mu(c)|^{2/(1+K/a)} \leq \zeta_a & \text{if } a \geq K \\[3mm] \sum_{c \varepsilon F_\tau} |\mu(c)| \leq \zeta_K & \text{if } a \leq K, \end{cases}$

where the positive constants ζ's above are obtained from (3.20) of Lemma 3.5. Finally, (5.18) and the definition of the p-variation of μ, given in (5.9), imply (5.16).   ⊠

Regarding the 'sharpness' of Theorem 5.6, we have only the following partial result:

<u>Theorem 5.7</u>

Let $F \subset [0,1]^J$ be infinite and suppose $\text{Dim}F = \text{dim}F$. Define p by (5.12).

If $\text{dim}F$ is exact then

$$(5.19) \qquad \sup\{|\mu|^q(F) : \mu \in F_{J/K}, \ \|\mu\|_{F_{J/K}} \leq 1\} = \infty$$

for all $q < p$.

If $\text{dim}F$ is asymptotic then

$$(5.20) \qquad \sup\{|\mu|^p(F) : \mu \in F_{J/K}, \ \|\mu\|_{F_{J/K}} \leq 1\} = \infty .$$

<u>Proof</u>

Let $F_o \subset F$ be a countable set whose combinatorial dimension equals $\text{dim}F$. Let Q_k be the kth canonical projection from $[0,1]^J$ onto $[0,1]$, and enumerate the countable set

$$Q_k[F_o] = \{x_k(i)\}_{i \in \mathbb{N}} \subset [0,1] , \quad k=1,\ldots,J .$$

By assumption, the combinatorial dimension of

$$\Lambda = \{(i_1,\ldots,i_J) : (x_1(i_1),\ldots,x_J(i_J)) \in F_o\}$$

equals $\text{dim}F_o = \text{dim}F$. To establish the theorem, it suffices to argue as follows: Suppose $d_\Lambda(b) = \infty$. By Lemma 3.6, for every $M > 0$ there exists a bounded J/K-linear form

$$f \sim \sum_{j \in \Lambda} a_j r_{P_1(j)} \otimes \cdots \otimes r_{P_{(\frac{J}{K})}(j)} ,$$

where $\{j \in \Lambda: a_j \neq 0\}$ is finite,

$$(5.21) \qquad \|f\|_\infty \leq 1 ,$$

and

$$(5.22) \qquad\qquad \sum_{j \in \Lambda} |a_j|^{2/(1+K/b)} > M .$$

Define a function μ on the generalized rectangles in $[0,1]^J$ by

$$\mu(P_1^{-1}[A_1] \cap \cdots \cap P_{\binom{J}{K}}^{-1}[A_{\binom{J}{K}}]) = \sum_{\substack{x \in P_1^{-1}[A_1] \cap \cdots \cap P_{\binom{J}{K}}^{-1}[A_{\binom{K}{K}}] \\ x=(x_1(j_1),\ldots,x_J(j_J)) \in F_o}} a_{j_1 \ldots j_J} ,$$

where $A_1,\ldots,A_{\binom{J}{K}}$ are Borel subsets of $[0,1]^K$. We conclude from (5.21) that $\mu \in F_{J/K}([0,1]^J)$ satisfies $\|\mu\|_{F_{J/K}} \leq 1$, and from (5.22) that it satisfies

$$|\mu|^{2/(1+K/b)}(F_o) > M .$$

We thus establish, since $M > 0$ was arbitrary, (5.19) and (5.20) in the statement of the theorem.   ☒

REFERENCES

1. P. Billingsley, <u>Ergodic Theory and Information</u>, John Wiley, New York, 1965.

2. R. C. Blei, Multidimensional extensions of the Grothendieck inequality and applications, Arkiv för Matematik, Vol. 17 (1979), No. 1, 51-68.

3. _____, Fractional Cartesian products of sets, Ann. Inst. Fourier, Grenoble 29, 2 (1979), 79-105.

4. _____, Combinatorial dimension and certain norms in harmonic analysis, Amer. J. of Math., Vol. 106 (1984), 847-887.

5. _____, and T. W. Körner, Combinatorial dimension and random sets, ISRAEL J. of Math., Vol. 47 (1984), 65-74.

6. M. Fréchet, Sur les fonctionnelles bilinéaires, Trans. Amer. Math. Soc., Vol. 16 (1915), 215-234.

7. U. Haagerup, Les Meilleures constantes de l'inégalité de Khintchine, C. R. Acad. Sc. Paris, t 286 (1978), A 259-262.

8. G. W. Johnson and G. S. Woodward, On p-Sidon sets, Indiana Univ. Math. J., 24 (1974), 161-167.

9. J. -P. Kahane, <u>Some Random Series of Functions</u>, Heath Math. Monographs, Mass., 1968.

10. Khintchine, J. Uber dyadische Bruche, Math. Zeit., 18 (1923), 109-116.

11. A. N. Kolmogorov and V. M. Tihomirov, ε-entropy and ε-capacity of sets in function spaces (in Russian), Usp. Mat. Nauk 14 (1959), 1-86; (English translation) American Math. Soc. Translations 17 (1961), 277-364.

12. J. E. Littlewood, On bounded bilinear forms in an infinite number of variables, Quart. J. Math. Oxford, 1 (1930), 164-174.

13. L. H. Loomis and H. Whitney, An inequality related to the isoperimetric inequality, Bulletin of A.M.S., Vol. 55, 9 (1949), 961-962.

14. M. Morse, Bimeasures and their integral extensions, Ann. Mat. Pura Appl., (4) 39 (1955), 345-356.

15. M. Morse and W. Transue, Functionals of bounded Fréchet variation, Canadian J. of Math, Vol. 1 (1949), 153-165.

16. R. Osserman, The isoperimetric inequality, Bulletin of A.M.S., Vol. 84, 6 (1978), 1182-1238.

17. G. Pisier, Sur l'espace des series de Fourier aléatoires presque surement continues, Exposé n°17-18, Seminaire sur la géométrie des espaces de Banach, Ecole Polytechnique, Palaiseau, 1977/78.

18. F. Riesz, Sur certains systemes singuliers d'équations intégrales, Annales Ecole Norm. Sup., (3) 28 (1911), 33-62.

19. W. Rudin, <u>Real and Complex Analysis</u>, McGraw-Hill, 1974.

20. E. Stein, <u>Singular Integrals and Differentiability Properties of Functions</u>, Princeton, New Jersey, 1970.

21. St. J. Szarek, On the best constant in the Khintchin inequality, Studia Math., 58 (1976), 197-208.

22. N. Th. Varopoulos, Tensor algebras and harmonic analysis, Acta Math., 119 (1967), 51-112.

Ron C. Blei
Department of Mathematics
The University of Connecticut
Storrs, CT   06268
USA

General instructions to authors for
PREPARING REPRODUCTION COPY FOR MEMOIRS

> For more detailed instructions send for AMS booklet, "A Guide for Authors of Memoirs."
> Write to Editorial Offices, American Mathematical Society, P. O. Box 6248,
> Providence, R. I. 02940.

MEMOIRS are printed by photo-offset from camera copy fully prepared by the author. This means that, except for a reduction in size of 20 to 30%, the finished book will look exactly like the copy submitted. Thus the author will want to use a good quality typewriter with a new, medium-inked black ribbon, and submit clean copy on the appropriate model paper.

Model Paper, provided at no cost by the AMS, is paper marked with blue lines that confine the copy to the appropriate size. Author should specify, when ordering, whether typewriter to be used has PICA-size (10 characters to the inch) or ELITE-size type (12 characters to the inch).

Line Spacing – For best appearance, and economy, a typewriter equipped with a half-space ratchet – 12 notches to the inch – should be used. (This may be purchased and attached at small cost.) Three notches make the desired spacing, which is equivalent to 1-1/2 ordinary single spaces. Where copy has a great many subscripts and superscripts, however, double spacing should be used.

Special Characters may be filled in carefully freehand, using dense black ink, or INSTANT ("rub-on") LETTERING may be used. AMS has a sheet of several hundred most-used symbols and letters which may be purchased for $5.

Diagrams may be drawn in black ink either directly on the model sheet, or on a separate sheet and pasted with rubber cement into spaces left for them in the text. Ballpoint pen is *not* acceptable.

Page Headings (Running Heads) should be centered, in CAPITAL LETTERS (preferably), at the top of the page – just above the blue line and touching it.

LEFT-hand, EVEN-numbered pages should be headed with the AUTHOR'S NAME;
RIGHT-hand, ODD-numbered pages should be headed with the TITLE of the paper (in shortened form if necessary).
Exceptions: PAGE 1 and any other page that carries a display title require NO RUNNING HEADS.

Page Numbers should be at the top of the page, on the same line with the running heads.

LEFT-hand, EVEN numbers – flush with left margin;
RIGHT-hand, ODD numbers – flush with right margin.
Exceptions: PAGE 1 and any other page that carries a display title should have page number, centered below the text, on blue line provided.

FRONT MATTER PAGES should be numbered with Roman numerals (lower case), positioned below text in same manner as described above.

MEMOIRS FORMAT

> It is suggested that the material be arranged in pages as indicated below.
> Note: <u>Starred items (*) are requirements of publication.</u>

Front Matter (first pages in book, preceding main body of text).

Page i – *Title, *Author's name.

Page iii – Table of contents.

Page iv – *Abstract (at least 1 sentence and at most 300 words).

*1980 Mathematics Subject Classification (1985 Revision). This classification represents the primary and secondary subjects of the paper, and the scheme can be found in Annual Subject Indexes of MATHEMATICAL REVIEWS beginning in 1984.

Key words and phrases, if desired. (A list which covers the content of the paper adequately enough to be useful for an information retrieval system.)

Page v, etc. – Preface, introduction, or any other matter not belonging in body of text.

Page 1 – Chapter Title (dropped 1 inch from top line, and centered).

Beginning of Text.

Footnotes: *Received by the editor date.
Support information – grants, credits, etc.

Last Page (at bottom) – Author's affiliation.

ABCDEFGHIJ – AMS – 898765